TEACHER RESOURCES

Earth History

Full Option Science System
Developed at the Lawrence Hall of Science, University of California, Berkeley
Published and Distributed by Delta Education

FOSS Lawrence Hall of Science Team

Larry Malone and Linda De Lucchi, FOSS Project Codirectors and Lead Developers; Jessica Penchos, Middle School Coordinator; Kathy Long, FOSS Assessment Director; David Lippman, Program Manager; Carol Sevilla, Publications Design Coordinator; Susan Stanley, Graphic Production; Rose Craig, Illustrator

FOSS Curriculum Developers: Alan Gould, Teri Lawson, Ann Moriarty, Virginia Reid, Joanna Snyder

FOSS Technology Developers: Susan Ketchner, Arzu Orgad

FOSS Multimedia Team: Kate Jordan, Senior Multimedia; Christopher Keller, Multimedia Producer; Jonathan Segal, Designer; Christopher Cianciarulo, Designer; Dan Bluestein, Programmer; Shan Jiang, Programmer

Delta Education Team

Bonnie A. Piotrowski, Editorial Director, Elementary Science

Project Team: Jennifer Apt, Sandra Burke, Mary Connell, Joann Hoy, Angela Miccinello, Jennifer McKenna

Content Reviewer

Gillian Puttick, TERC

Thank you to all FOSS Middle School Revision Trial Teachers and District Coordinators

Frances Amojioyi, Lincoln Middle School, Alameda, CA; Dean Anderson, Organized trials for Boston Public Schools, Boston, MA; Thomas Archer, Organized trials for ESD 112, Vancouver, WA; Lauresa Baker, Lincoln Middle School, Alameda, CA; Bobbi Anne Barnowsky, Canyon Middle School, Castro Valley, CA; Christine Bertko, St. Finn Barr Catholic School, San Francisco, CA; Stephanie Billinge, James P. Timilty Middle School, Roxbury, MA; Jerry Breton, Ingleside Middle School, Phoenix, AZ; Robert Cho, Timilty Middle School, Boston, MA; Susan Cohen, Cherokee Heights Middle School, Madison, WI; Malcolm Davis, Canyon Middle School, Castro Valley, CA; Marilyn Decker, Organized trials for Milton PS, Milton, MA; Jenny Ernst, Park Day School, Oakland, CA; Marianne Floyd, Spanaway Middle School, Spanaway, WA; Sarah Kathryn Gessford, Journeys School, Jackson, WY; Charles Hardin, Prairie Point Middle School, Cedar Rapids, IA; Jennifer Hartigan, Lincoln Middle School, Alameda, CA; Sheila Holland, TechBoston Academy, Boston, MA; Nicole Hoyceanyls, Charles S. Pierce Middle School, Milton, MA; Bruce Kamerer, Donald McKay K-8 School, East Boston, MA; Carmen Saele Kardokus, Reeves Middle School, Olympia, WA; Janey Kaufman, Organized trials for Scottsdale USD, Scottsdale, AZ; Erica Larson, Organized trials for Grant Wood AEA, Cedar Rapids, IA; Lindsay Lodholz, O'Keeffe Middle School, Madison, WI; Robert Mattisinko, Chaparral High School, Scottsdale, AZ; Brenda McGurk, Prairie Point Middle School, Cedar Rapids, IA; Tim Miller, Mountainside Middle School, Scottsdale, AZ; Thomas Miro, Lincoln Middle School, Alameda, CA; Spencer Nedved, Frontier Middle School, Vancouver, WA; Joslyn Olsen, Lincoln Middle School, Alameda, CA; Stephanie Ovechka, Cedarcrest Middle School, Spanaway, WA; Barbara Reinert, Copper Ridge School, Scottsdale, AZ; Stephen Ramos, Lincoln Middle School, Alameda, CA; Gina Rutenbeck, Prairie Point Middle School, Cedar Rapids, IA; John Sheridan, Boston Public Schools (Boston Schoolyard Initiative), Boston, MA; Barbara Simon, Timilty Middle School, Boston, MA; Lise Simpson, Alcott Middle School, Norman, OK; Autumn Stevick, Thurgood Marshall Middle School, Olympia, WA; Ted Stoeckley, Hall Middle School, Larkspur, CA; Lesli Taschwer, Organized trials for Madison SD, Madison, WI; Paula Warner, Alcott Middle School, Norman, OK; Darren T. Wells, James P. Timilty Middle School, Boston, MA; Kristin White, Frontier Middle School, Vancouver, WA

Photo Credits: © ronnybas/Shutterstock (cover); © Delta Education

Published and Distributed by Delta Education, a member of the School Specialty Family

The FOSS program was developed in part with the support of the National Science Foundation grant nos. ESI-9553600 and ESI-0242510. However, any opinions, findings, conclusions, statements, and recommendations expressed herein are those of the authors and do not necessarily reflect the views of NSF. FOSSmap was developed in collaboration between the BEAR Center at UC Berkeley and FOSS at the Lawrence Hall of Science. Score analysis is done through the BEAR Center Scoring Engine.

Earth History — Teacher Toolkit, 1558522
Teacher Resources, 1558537
978-1-62571-806-8
Printing 2 – 1/2019
Patterson Printing, Benton Harbor, MI

WARNING — This set contains chemicals that may be harmful if misused. Read cautions on individual containers carefully. Not to be used by children except under adult supervision.

This warning label is required by the
U.S. Consumer Product Safety Commission.
The chemicals in the FOSS Earth History kit are
calcium hydroxide, phenyl salicylate, and hydrochloric acid.

TABLE OF CONTENTS

This document, *Teacher Resources*, is one of three parts of the *FOSS Teacher Toolkit* for this course. The chapters in *Teacher Resources* are all available as PDFs on FOSSweb.

The other parts of the course *Teacher Toolkit* are the *Investigations Guide* and a copy of the *FOSS Science Resources* student book containing original readings for this course.

The spiral-bound *Investigations Guide* contains these chapters.

- Overview
- Framework and NGSS
- Materials
- Investigations
- Assessment

The *Teacher Toolkit* is the most important part of the FOSS Program. It is here that all the wisdom and experience contributed by hundreds of educators has been assembled. Everything we know about the content of the course, how to teach the subject, and the resources that will assist the effort are presented here.

FOSS Program Goals

FOSS Program Goals

Contents

INTRODUCTION

The Full Option Science System™ has evolved from a philosophy of teaching and learning at the Lawrence Hall of Science that has guided the development of successful active-learning science curricula for more than 40 years. The FOSS Program bridges research and practice by providing tools and strategies to engage students and teachers in enduring experiences that lead to deeper understanding of the natural and designed worlds.

Science is a creative and analytic enterprise, made active by our human capacity to think. Scientific knowledge advances when scientists observe objects and events, think about how they relate to what is known, test their ideas in logical ways, and generate explanations that integrate the new information into understanding of the natural world. Engineers apply that understanding to solve real-world problems. Thus the scientific enterprise is both what we know (content knowledge) and how we come to know it (practices). Science is a discovery activity, a process for producing new knowledge.

The best way for students to appreciate the scientific enterprise, learn important scientific and engineering concepts, and develop the ability to think well is to actively participate in scientific practices through their own investigations and analyses. FOSS was created to engage students and teachers with meaningful experiences in the natural and designed worlds.

GOALS OF THE FOSS PROGRAM

FOSS has set out to achieve three important goals: scientific literacy, instructional efficiency, and systemic reform.

Scientific Literacy

FOSS provides all students with science experiences that are appropriate to students' cognitive development and prior experiences. It provides a foundation for more advanced understanding of core science ideas that are organized in thoughtfully designed learning progressions and prepares students for life in an increasingly complex scientific and technological world.

The National Research Council (NRC) in *A Framework for K–12 Science Education: Practices, Crosscutting Concepts, and Core Ideas* and the American Association for the Advancement of Science (AAAS) in *Benchmarks for Scientific Literacy* have described the characteristics of scientific literacy:

- Familiarity with the natural world, its diversity, and its interdependence.
- Understanding the disciplinary core ideas and the crosscutting concepts of science, such as patterns; cause and effect; scale, proportion, and quantity; systems and system models; energy and matter—flows, cycles, and conservation; structure and function; and stability and change.
- Knowing that science and engineering, technology, and mathematics are interdependent human enterprises and, as such, have implied strengths and limitations.
- Ability to reason scientifically.
- Using scientific knowledge and scientific and engineering practices for personal and social purposes.

The FOSS Program design is based on learning progressions that provide students with opportunities to investigate core ideas in science in increasingly complex ways over time. FOSS starts with the intuitive ideas that primary students bring with them and provides experiences that allow students to develop more sophisticated understanding as they grow through the grades. Cognitive research tells us that learning involves individuals in actively constructing schemata to organize new information and to relate and incorporate the new understanding into established knowledge. What sets experts apart from novices is that

experts in a discipline have extensive knowledge that is effectively organized into structured schemata to promote thinking. Novices have disconnected ideas about a topic that are difficult to retrieve and use. Through internal processes to establish schemata and through social processes of interacting with peers and adults, students construct understanding of the natural world and their relationship to it.

The target goal for FOSS students is to know and use scientific explanations of the natural world and the designed world; to understand the nature and development of scientific knowledge and technological capabilities; and to participate productively in scientific and engineering practices.

Instructional Efficiency

FOSS provides all teachers with a complete, cohesive, flexible, easy-to-use science program that reflects current research on teaching and learning, including student discourse, argumentation, writing to learn, and reflective thinking, as well as teacher use of formative assessment to guide instruction. The FOSS Program uses effective instructional methodologies, including active learning, scientific practices, focus questions to guide inquiry, working in collaborative groups, multisensory strategies, integration of literacy, appropriate use of digital technologies, and making connections to students' lives.

FOSS is designed to make active learning in science engaging for teachers as well as for students. It includes these supports for teachers:

- Complete equipment kits with durable, well-designed materials for all students.
- Detailed *Investigations Guid*e with science background for the teacher and focus questions to guide instructional practice and student thinking.
- Multiple strategies for formative assessment.
- Benchmark assessments with scoring guides.
- Strategies for use of science notebooks for novice and experienced users.
- *FOSS Science Resources*, a book of course-specific readings.
- The FOSS website with course-integrated online activities for use in school or at home, suggested extension activities, and extensive online support for teachers.

FOSS Program Goals

Systemic Reform

FOSS provides schools and school systems with a program that addresses the community science-achievement standards. The FOSS Program prepares students by helping them acquire the knowledge and thinking capacity appropriate for world citizens.

The FOSS Program design makes it appropriate for reform efforts on all scales. It reflects the core ideas to be incorporated into the next-generation science standards. It meets with the approval of science and technology companies working in collaboration with school systems, and it has demonstrated its effectiveness with diverse student and teacher populations in major urban reform efforts. The use of science notebooks and formative-assessment strategies in FOSS redefines the role of science in a school—the way that teachers engage in science teaching with one another as professionals and with students as learners, and the way that students engage in science learning with the teacher and with one another. FOSS takes students and teachers beyond the classroom walls to establish larger communities of learners.

BRIDGING RESEARCH INTO PRACTICE

The FOSS Program is built on the assumptions that understanding core scientific knowledge and how science functions is essential for citizenship, that all teachers can teach science, and that all students can learn science. The guiding principles of the FOSS design, described below, are derived from research and confirmed through FOSS developers' extensive experience with teachers and students in typical American classrooms.

Understanding of science develops over time. FOSS has elaborated learning or content progressions for core ideas in science for kindergarten through grade 8. Developing the learning progressions involves identifying successively more sophisticated ways of thinking about core ideas over multiple years. "If mastery of a core idea in a science discipline is the ultimate educational destination, then well-designed learning progressions provide a map of the routes that can be taken to reach that destination" (National Research Council, *A Framework for K–12 Science Education*, 2011).

Focusing on a limited number of topics in science avoids shallow coverage and provides more time to explore core science ideas in depth. Research emphasizes that fewer topics experienced in greater depth produces much better learning than many topics briefly visited. FOSS affirms this research. FOSS courses provide long-term engagement (10–12 weeks) with important science ideas. Furthermore, courses build upon one another within and across each strand, progressively moving students toward the grand ideas of science. The core ideas of science are difficult and complex, never learned in one lesson or in one class year.

FOSS Next Generation—K–8 Sequence

	PHYSICAL SCIENCE		EARTH SCIENCE		LIFE SCIENCE	
	MATTER	ENERGY AND CHANGE	ATMOSPHERE AND EARTH	ROCKS AND LANDFORMS	STRUCTURE/ FUNCTION	COMPLEX SYSTEMS
6–8	Waves; Gravity and Kinetic Energy Chemical Interactions Electromagnetic Force; Variables and Design		Planetary Science Earth History Weather and Water		Heredity and Adaptation; Human Systems Interactions Populations and Ecosystems Diversity of Life	
5	Mixtures and Solutions		Earth and Sun		Living Systems	
4		Energy		Soils, Rocks, and Landforms	Environments	
3	Motion and Matter		Water and Climate		Structures of Life	
2	Solids and Liquids			Pebbles, Sand, and Silt	Insects and Plants	
1		Sound and Light	Air and Weather		Plants and Animals	
K	Materials and Motion		Trees and Weather		Animals Two by Two	

FOSS Program Goals

Science is more than a body of knowledge. How well you think is often more important than how much you know. In addition to the science content framework, every FOSS course provides opportunities for students to engage in and understand science practices, and many courses explore issues related to engineering practices and the use of natural resources. FOSS uses these science and engineering practices.

- Asking questions (for science) and defining problems (for engineering)
- Developing and using models
- Planning and carrying out investigations
- Analyzing and interpreting data
- Using mathematics, information and computer technology, and computational thinking
- Constructing explanations (for science) and designing solutions (for engineering)
- Engaging in argument from evidence
- Obtaining, evaluating, and communicating information

Science is inherently interesting, and children are natural investigators. It is widely accepted that children learn science concepts best by doing science. Doing science means hands-on experiences with objects, organisms, and systems. Hands-on activities are motivating for students, and they stimulate inquiry and curiosity. For these reasons, FOSS is committed to providing the best possible materials and the most effective procedures for deeply engaging students with scientific concepts. FOSS students at all grade levels investigate, experiment, gather data, organize results, and draw conclusions based on their own actions. The information gathered in such activities enhances the development of science and engineering practices.

Education is an adventure in self-discovery. Science provides the opportunity to connect to students' interests and experiences. Prior experiences and individual learning styles are important considerations for developing understanding. Observing is often equated with seeing, but in the FOSS Program all senses are used to promote greater understanding. FOSS evolved from pioneering work done in the 1970s with students with disabilities. The legacy of that work is that FOSS investigations naturally use multisensory methods to accommodate students with physical and learning disabilities and also to maximize information gathering for all students. A number of tools, such as the FOSS syringe and balance, were originally designed to serve the needs of students with disabilities.

Formative assessment is a powerful tool to promote learning and can change the culture of the learning environment. Formative assessment in FOSS creates a community of reflective practice. Teachers and students make up the community and establish norms of mutual support, trust, respect, and collaboration. The goal of the community is that everyone will demonstrate progress and will learn and grow.

Science-centered language development promotes learning in all areas. Effective use of science notebooks can promote reflective thinking and contribute to lifelong learning. Research has shown that when language-arts experiences are embedded within the context of learning science, students improve in their ability to use their language skills. Students are motivated to read to find out information, and to share their experiences both verbally and in writing.

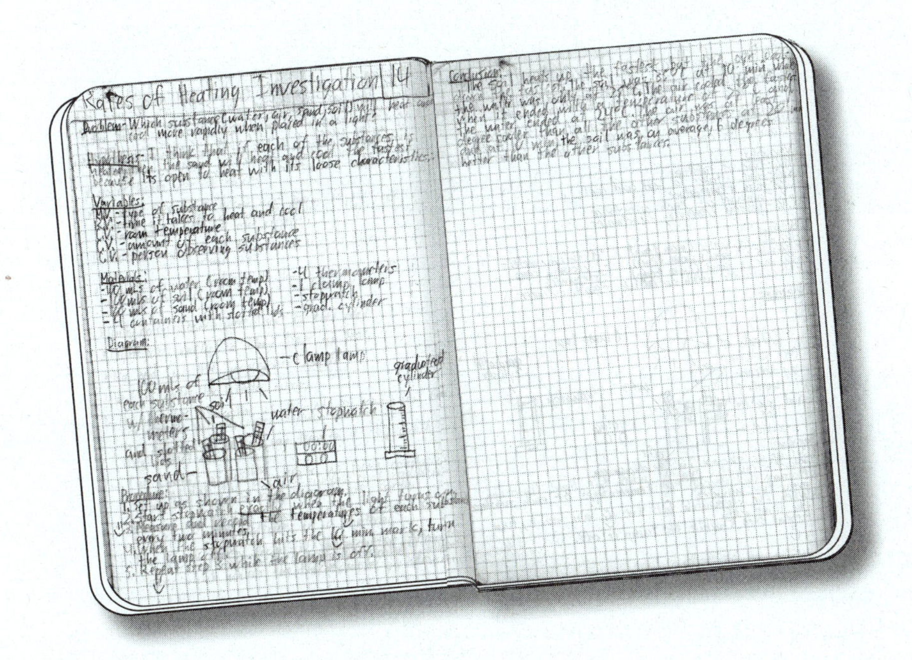

Experiences out of the classroom develop awareness of community. By extending classroom learning into the local region and community, FOSS brings the science concepts and principles to life. In the process of extending classroom learning to the natural world and utilizing community resources, students will develop a relationship with learning that extends beyond the classroom walls.

FOSS NEXT GENERATION K–8 SCOPE AND SEQUENCE

Grade	Physical Science	Earth Science	Life Science
6–8	Waves* Gravity and Kinetic Energy*	Planetary Science	Heredity and Adaptation* Human Systems Interactions*
	Chemical Interactions	Earth History	Populations and Ecosystems
	Electromagnetic Force* Variables and Design*	Weather and Water	Diversity of Life
5	Mixtures and Solutions	Earth and Sun	Living Systems
4	Energy	Soils, Rocks, and Landforms	Environments
3	Motion and Matter	Water and Climate	Structures of Life
2	Solids and Liquids	Pebbles, Sand, and Silt	Insects and Plants
1	Sound and Light	Air and Weather	Plants and Animals
K	Materials and Motion	Trees and Weather	Animals Two by Two

*** Half-length course**

FOSS is a research-based science curriculum for grades K–8 developed at the Lawrence Hall of Science, University of California, Berkeley. FOSS is also an ongoing research project dedicated to improving the learning and teaching of science. The FOSS project began over 25 years ago during a time of growing concern that our nation was not providing young students with an adequate science education. The FOSS Program materials are designed to meet the challenge of providing meaningful science education for all students in diverse American classrooms and to prepare them for life in the 21st century. Development of the FOSS Program was, and continues to be, guided by advances in the understanding of how people think and learn.

With the initial support of the National Science Foundation and continued support from the University of California, Berkeley, and School Specialty, Inc., the FOSS Program has evolved into a curriculum for all students and their teachers, grades K–8. The current editions of FOSS are the result of a rich collaboration among the FOSS/Lawrence Hall of Science development staff; the FOSS product development team at School Specialty; assessment specialists, educational researchers, and scientists; and dedicated professionals in the classroom and their students, administrators, and families.

We acknowledge the thousands of FOSS educators who have embraced the notion that science is an active process, and we thank them for their significant contributions to the development and implementation of the FOSS Program.

Science Notebooks in Middle School

Science Notebooks in Middle School

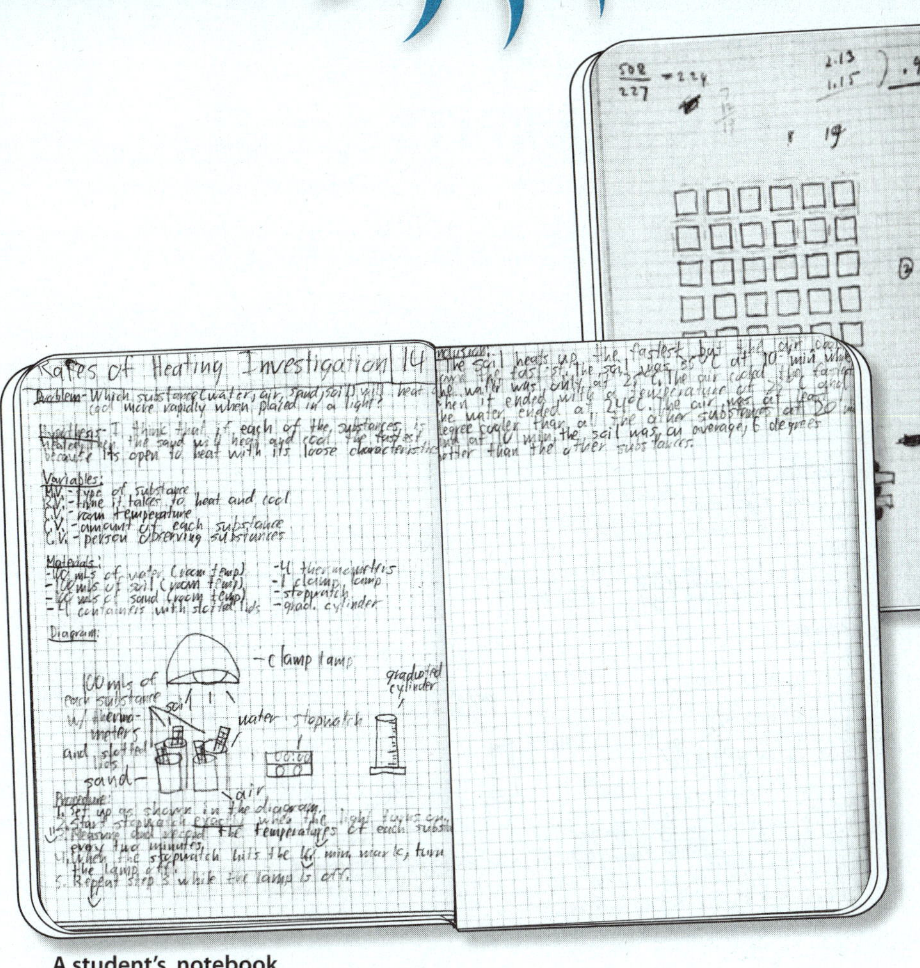

A scientist's notebook

Source: *Special Collections Research Center, University of Chicago Library*

A student's notebook

INTRODUCTION

Scientists keep notebooks. The scientist's notebook is a detailed record of his or her engagement with scientific phenomena. It is a personal representation of experiences, observations, and thinking—an integral part of the process of doing scientific work. A scientist's notebook is a continuously updated history of the development of scientific knowledge and reasoning. The notebook organizes the huge body of knowledge and makes it easier for a scientist to work. As developing scientists, FOSS students are encouraged to incorporate notebooks into their science learning. First and foremost, the notebook is a tool for student learning.

Contents

NOTEBOOK BENEFITS

Engaging in active science is one part experience and two parts making sense of the experience. Science notebooks help students with the sense-making part by providing two major benefits: documentation and cognitive engagement.

From the Human Systems Interactions Course

Benefits to Students

Science notebooks centralize students' data. When data are displayed in functional ways, students can think about the data more effectively. A well-kept notebook is a useful reference document. When students have forgotten a fact or relationship that they learned earlier in their studies, they can look it up. Learning to reference previous discoveries and knowledge structures is important.

Documentation: an organized record. As students become more accomplished at keeping notebooks, their work will become better organized and efficient. Tables, graphs, charts, drawings, and labeled illustrations will become standard means for representing and displaying data. A complete and accurate record of learning allows students to reconstruct the sequence of learning events and relive the experience. Discussions about science among students, students and teachers, or students, teachers, and families, have more meaning when they are supported by authentic documentation in students' notebooks. Questions and ideas generated by experimentation or discussion can be recorded for future investigation.

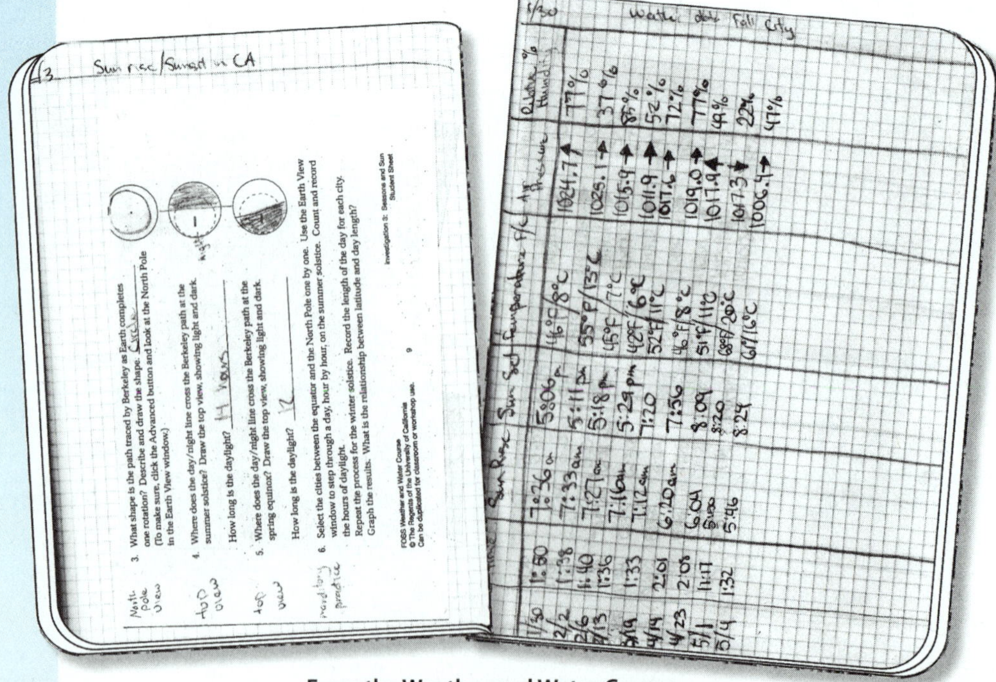

From the Weather and Water Course

Cognitive engagement. Once data are recorded and organized in an efficient manner in science notebooks, students can think about the data and draw conclusions about the way the world works. Their data are the raw materials that students use to forge concepts and relationships from their experiences and observations.

Writing stimulates active reasoning. There is a direct relationship between the formation of concepts and the rigors of expressing them in words. Writing requires students to impose discipline on their thoughts. When you ask students to generate derivative products (summary reports, detailed explanations, posters, oral presentations, etc.) as evidence of learning, the process will be much more efficient and meaningful because they have a coherent, detailed notebook for reference.

When students use notebooks as an integral part of their science studies, they think critically about their thinking. This reflective thinking can be encouraged by notebook entries that present opportunities for self-assessment. Self-assessment motivates students to rethink and restate their scientific understanding. Revising their notebook entries helps students clarify their understanding of the science concepts under investigation. By writing explanations, students clarify what they know and expose what they don't know.

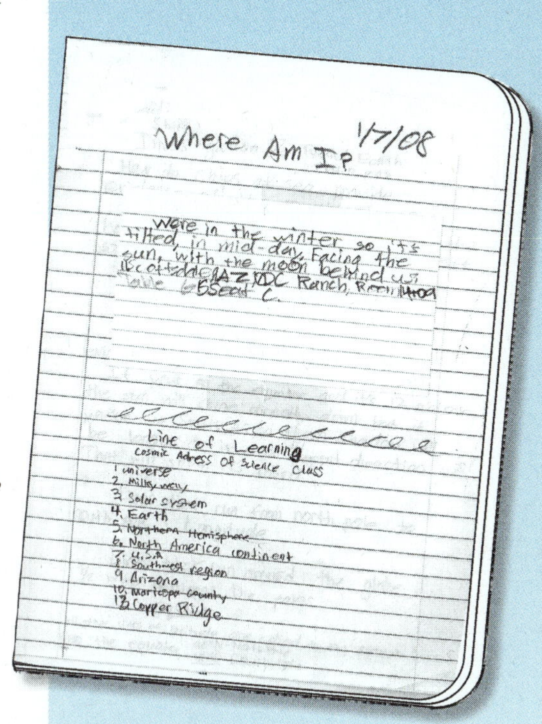

From the Planetary Science Course

11/30/11

Calculating Asteroid Size

Directions:

Use the 20:1 crater-to-asteroid ratio to determine the size of the asteroid that created these Earth craters.

Earth Crater	Diameter (km)	Asteroid Size (km)
Manicouagan (p. 76)	100	5
Chicxulub (p. 33)	300	15
Gosses Bluff (p. 75)	24	1.2
Barringer (p. 74)	1.2	0.06

Lunar Crater	Diameter (km)	Asteroid Size (km)
Archimedes (p. 70)	83	4.1
Aristillus (p. 70)	55	2.8
Clavius (p. 73)	225	11.3
Copernicus (p. 71)	93	4.6
Posidonius (p. 72)	95	4.8
Stofler (p. 73)	126	6.3
Tycho (p. 73)	85	4.2
Sea of Rains (p. 65)	1000	50

Comparing asteroid impacts

Why are there more craters on the Moon than on Earth?

On Earth, craters get covered by plants, buildings, and water. They can also dissapear from weather like rain and wind.

From the Planetary Science Course

Benefits to Teachers

In FOSS, the unit of instruction is the course—a sequence of conceptually related learning experiences that leads to a set of learning outcomes. A science notebook helps you think about and communicate the conceptual structure of the course you are teaching.

Assessment. From the assessment point of view, a science notebook is a collection of student-generated artifacts that exhibit learning. You can informally assess student skills, such as using charts to record data, in real time while students are working with materials. At other times, you might collect student work samples and review them for insights or errors in conceptual understanding. This valuable information helps you plan the next steps of instruction. Students' data analysis, sense making, and reflection provide a measure of the quality and quantity of student learning. The notebook itself should not be graded, though certain assignments might be graded and placed in the notebook.

Medium for feedback. The science notebook provides an excellent medium for providing feedback to individual students regarding their work. Productive feedback calls for students to read a teacher comment, think about the issue it raises, and act on it. The comment may ask for clarification, an example, additional information, precise vocabulary, or a review of previous work in the notebook. In this way, you can determine whether a problem with the student work relates to a flawed understanding of the science content or a breakdown in communication skills.

Focus for professional discussions. The student notebook also acts as a focal point for discussion about student learning at several levels. First, a student's work can be the subject of a conversation between you and the student. By acting as a critical mentor, you can call attention to ways a student can improve the notebook, and help him or her learn how to use the notebook as a reference. You can also review and discuss the science notebook during family conferences. Science notebooks shared among teachers in a study group or other professional-development environment can effectively demonstrate recording techniques, individual styles, various levels of work quality, and so on. Just as students can learn notebook strategies from one another, teachers can learn notebook skills from one another.

GETTING STARTED

A middle school science notebook is more than just a collection of science work, notes, field-trip permission slips, and all the other types of documents that tend to accumulate in a student's three-ring binder or backpack. By organizing the science work systematically into a bound composition book, students create a thematic record of their experiences, thoughts, plans, reflections, and questions as they work through a topic in science.

The science notebook is more than just formal lab reports; it is a record of a student's entire journey through a progression of science concepts. Where elementary school students typically need additional help structuring and organizing their written work, middle school students should be encouraged to develop their organizational skills and take some ownership in creating deliberate records of their science learning, even though they may still require some pointers and specific scaffolding from you.

In addition, the science notebook provides a personal space where students can explore their understanding of science concepts by writing down ideas and being allowed to "mess around" with their thinking. Students are encouraged to look back on their ideas throughout the course to self-assess their conceptual development and record new thoughts. With this purpose of the science notebook in mind, you may need to refine your own thinking around what should or should not be included as a part of the science notebook, as well as expectations about grading and analyzing student work.

Rules of Engagement

Teachers and students should be clear about the conventions students will honor in their notebook entries. Typically, the rules of grammar and spelling are fairly relaxed so as not to inhibit the flow of expression during notebook entries. This also helps students develop a sense of ownership in their notebooks, a place where they are free to write in their own style. When students generate derivative products using information in the notebooks, such as reports, you might require students to exercise more rigorous language-arts conventions.

In addition to written entries, students should be encouraged to use a wide range of other means for recording and communicating, including charts, tables, graphs, drawings, graphics, color codes, numbers, and artifacts attached to the notebook pages. By expanding the options for making notebook entries, each student will find his or her most efficient, expressive way to capture and organize information for later retrieval.

Enhanced Classroom Discussion

One of the benefits of using notebooks is that you will elicit responses to key discussion questions from all students, not just the handful of verbally enthusiastic students in the class. When you ask students to write down their thoughts after you pose a question, all students have time to engage deeply with the question and organize their thoughts. When you ask students to share their answers, those who needed more time to process the question and organize their thinking will be ready to verbalize their responses and become involved in a class discussion.

When students can use their notebooks as a reference during the ensuing discussion, they won't feel put on the spot. At some points, you might ask students to share only what they wrote in their notebooks, to remind them to focus their thoughts while writing. As the class shares ideas during discussions, students can add new ideas to their notebooks under a line of learning (see next-step strategies). Even if some students are still reticent, having students write after a question is posed prevents them from automatically disengaging from conversations.

Notebook Structure

FOSS recommends that students keep their notebooks in 8" × 10" bound composition books. At the most advanced level, students are responsible for creating the entire science notebook from blank pages in their composition books. Experienced students determine when to use their notebooks, how to organize space, what methods of documentation to use, and how to flag important information. This level of notebook use will not be realized quickly; it will likely require systematic development by an entire teaching staff over time.

At the beginning, notebook practice is often highly structured, using prepared sheets from the FOSS notebook masters. You can photocopy and distribute these sheets to students as needed during the investigations. Sheets are sized to fit in a standard composition book. Students glue or tape the sheets into their notebooks. This allows some flexibility between glued-in notebook sheets and blank pages where students can do additional writing, drawings, and other documentation. Prepared notebook sheets are helpful organizers for students with challenges such as learning disabilities or with developing English skills. This model is the most efficient means for obtaining the most productive work from inexperienced middle school students.

From the Planetary Science Course

To make it easy for new FOSS teachers to implement a beginning student notebook, Delta Education sells copies of the printed *FOSS Science Notebook* in English for all FOSS middle school courses. Electronic versions of the notebook sheets can be downloaded free of charge at www.FOSSweb.com.

Each *FOSS Science Notebook* is a bound set of the notebook sheets for the course plus extra blank sheets throughout the notebook for students to write focus or inquiry questions, record and organize data, make sense of their thinking, and write summaries. There are also blank pages at the end to develop an index of science vocabulary.

The questions, statements, and graphic organizers on the notebook sheets provide guidance for students and scaffolding for teachers. When the notebook sheets are organized as a series, they constitute a highly structured precursor to an autonomously generated science notebook.

Developing Notebook Skills

Students will initially need more guidance from you. You will need to describe what and when to record, and to model organizational techniques. As the year advances, the notebook work will become increasingly student centered. As the body of work in the notebook grows, students will have more and more examples of useful techniques for reference. This self-sufficiency reduces the amount of guidance you need to provide, and reinforces students' appreciation of their own record of learning.

This gradual shift toward student-centered use of the notebook applies to any number of notebook skills, including developing headers for each page (day, time, date, title, etc.); using space efficiently on the page; preparing graphs, graphic organizers, and labeled illustrations; and attaching artifacts (sand samples, dried flowers, photographs, etc.). For instance, when students first display their data in a two-coordinate graph, the graph might be completely set up for them, so that they simply plot the data. As the year progresses, they will be expected to produce graphs with less and less support, until they are doing so without any assistance from you.

Organizing Science Notebooks

Four organizational components of the notebook should be planned right from the outset: a table of contents, page numbering, entry format, and an index.

Table of contents. Students should reserve the first three to five pages of their notebook for the table of contents. They will add to it systematically as they proceed through the course. The table of contents should include the date, title, and page number for each entry. The title could be based on the names of the investigations in the course, the specific activities undertaken, the concepts learned, a focus question for each investigation, or some other schema that makes sense to everyone.

Page numbering. Each page should have a number. These are referenced in the table of contents as the notebook progresses.

Entry format. During each class session, students will document their learning. Certain information will appear in every record, such as the date and title. Other forms of documentation will vary, including different types of written entries and artifacts, such as a multimedia printout. Some teachers ask their students to start each new entry at the top of the next available page. Others simply leave a modest space before a new entry. Sometimes it is necessary to leave space for work that will be completed on a separate piece of paper and glued or taped in later. Students might also leave space after a response, so that they can add to it at a later time.

Index. Scientific academic language is important. FOSS strives to have students use precise, accurate vocabulary at all times in their writing and conversations. To help them learn scientific vocabulary, students should set up an index at the end of their notebooks. It is not usually possible for students to enter the words in alphabetical order, as they will be acquired as the course advances. Instead, students could use several pages at the end of the notebook blocked out in 24 squares, and assign one or more letters to each square. Students write the new vocabulary word or phrase in the appropriate square and tag it with the page number of the notebook on which the word is defined. By developing vocabulary in context, students construct meaning through the inquiry process, and by organizing the words in an index, they strengthen their science notebooks as a documentary tool of their science learning. As another alternative, students can also define the word within these squares with the page references.

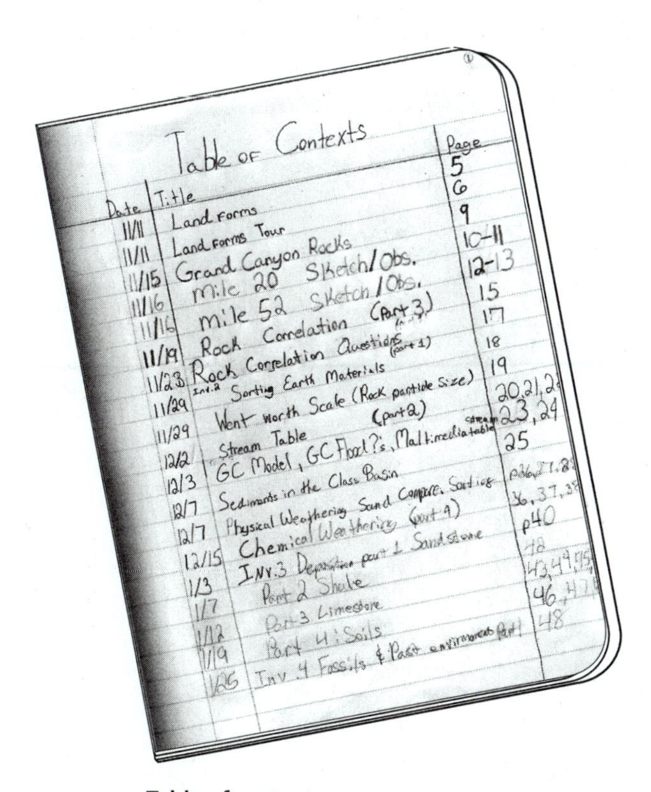

Table of contents

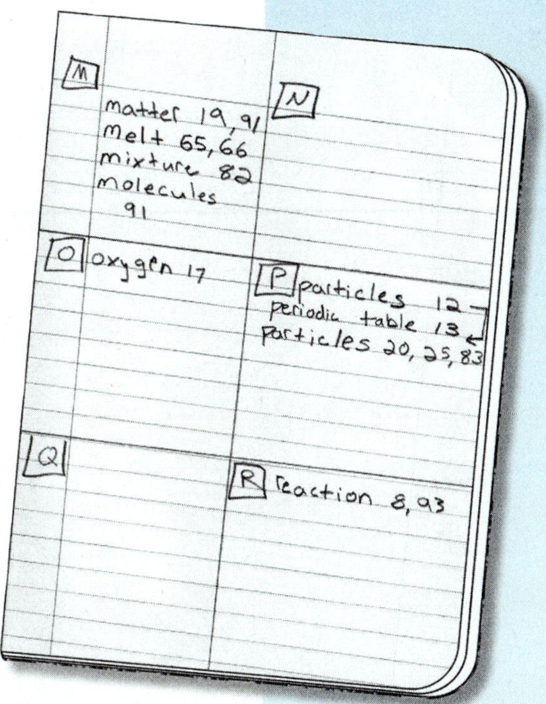

A science notebook index

NOTEBOOK COMPONENTS

Four general types of notebook entries, or components, give the science notebook conceptual shape and direction. These structures don't prescribe a step-by-step procedure for how to prepare the notebook, but they do provide some overall guidance. The general arc of an investigation starts with a question or challenge, proceeds with an activity and data acquisition, continues to sense making, and ends with next steps such as reflection and self-assessment.

All four components are not necessary during each class session, but over the course of an investigation, each component will be visited at least once. It may be useful to keep these four components in mind as you systematically guide students through their notebook entries. The components are overviewed here and described in greater detail on the following pages.

Focusing the investigation. Each part of each FOSS investigation includes a focus question, which students transcribe into their notebooks. Focus questions are embedded in the teacher step-by-step instructions and explicitly labeled. The focus question establishes the direction and conceptual challenge for that part of the investigation. For instance, when students investigate the origins of sand and sandstone in the **Earth History Course**, they start by writing,

➤ *Which came first, sand or sandstone?*

The question focuses both students and you on the learning goals for the activity. Students may start by formulating a plan, formally or informally, for answering the focus question. The goal of the plan is to obtain a satisfactory answer to the focus question, which will be revisited and answered later in the investigation.

Data acquisition and organization. After students have established a plan, they collect data. Students can acquire data from carefully planned experiments, accurate measurements, systematic observations, free explorations, or accidental discoveries. It doesn't matter what process produces the data; the critically important point is that students obtain data and record it. It may be necessary to reorganize and display the data for efficient analysis, often by organizing a data table. The data display is key to making sense of the science inquiry.

Making sense of data. Once students have collected and displayed their data, they need to analyze it to learn something about the natural world. In this component of the notebook, students write explanatory statements that answer the focus question. You can formalize this component by asking students to use an established protocol such as a sentence starter, or the explanation can be purely a thoughtful effort by each student. Explanations may be incorrect or incomplete at this point, but students can remedy this during the final notebook entry, when they have an opportunity to continue processing what they've learned. Unfortunately, this piece is often forgotten in the classroom during the rush to finish the lesson and move on. But without sense making and reflection (the final phase of science inquiry), students might see the lesson as a fun activity without connecting the experience to the big ideas that are being developed in the course.

Next-step strategies. The final component of an investigation brings students back to their notebooks by engaging in a next-step strategy, such as reflection and self-assessment, that moves their understanding forward. This component is the capstone on a purposeful series of experiences designed to guide students to understand the concept originally presented in the focus question. After making sense of the data, and making new claims about the topic at hand, students should go back to their earlier thinking and note their changing ideas and new findings. This reflective process helps students cement their new ideas.

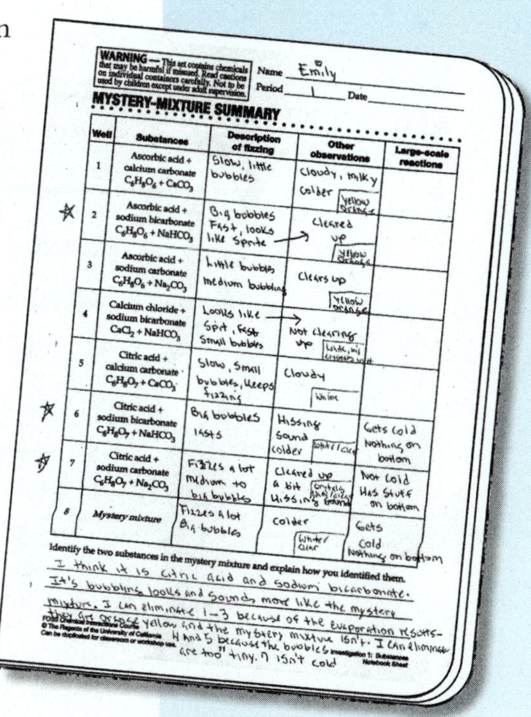

WARNING — This set contains chemicals that may be harmful if misused. Read cautions on individual containers carefully. Not to be used by children except under adult supervision.

Name Emily

Period 1 Date

MYSTERY-MIXTURE SUMMARY

Well	Substances	Description of fizzing	Other observations	Large-scale reactions
1	Ascorbic acid + calcium carbonate $C_6H_8O_6 + CaCO_3$	Slow, little bubbles	Cloudy, milky colder; yellow orange	
2	Ascorbic acid + sodium bicarbonate $C_6H_8O_6 + NaHCO_3$	Big bubbles Fast, looks like Sprite	Cleared up; yellow orange	
3	Ascorbic acid + sodium carbonate $C_6H_8O_6 + Na_2CO_3$	little bubbles medium bubbling	Clears up; yellow orange	
4	Calcium chloride + sodium bicarbonate $CaCl_2 + NaHCO_3$	Looks like Spit, fast Small bubbles	Not clearing up; white, big crystals	
5	Citric acid + calcium carbonate $C_6H_8O_7 + CaCO_3$	Slow, Small bubbles, keeps fizzing	Cloudy; white	
6	Citric acid + sodium bicarbonate $C_6H_8O_7 + NaHCO_3$	Big bubbles fasts	Hissing sound colder; white/clear	Gets cold Nothing on bottom
7	Citric acid + sodium carbonate $C_6H_8O_7 + Na_2CO_3$	Fizzes a lot medium to big bubbles	Cleared up a bit Hissing sound; crystals white/clear	Not cold Has stuff on bottom
8	*Mystery mixture*	Fizzes a lot Big bubbles	colder; white/clear	Gets Cold Nothing on bottom

Identify the two substances in the mystery mixture and explain how you identified them.

I think it is citric acid and sodium bicarbonate. It's bubbling looks and sounds more like the mystery mixture. I can eliminate 1–3 because of the evaporation results–they are orange yellow and the mystery mixture isn't. I can eliminate 4 and 5 because the bubbles are too tiny. 7 isn't cold

FOSS Chemical Interactions Course
© The Regents of the University of California
Can be duplicated for classroom or workshop use.

Investigation 1: Substances
Notebook Sheet

A student organizes and makes sense of data in the Chemical Interactions Course.

Focusing the Investigation

Focus question. The first notebook entry in most investigations is the focus question. Focus questions are embedded in the teacher step-by-step instructions and explicitly labeled. You can write the question on the board or project it for students to transcribe into their notebooks. The focus question serves to focus students and you on the inquiry for the day. It is not always answered immediately, but rather hangs in the air while the investigation goes forward. Students always revisit their initial responses later in the investigation.

Quick write. A quick write (or quick draw) can be used in addition to a focus question. Quick writes can be completed on a quarter sheet of paper or an index card so you can collect, review, and return them to students to be taped or glued into their notebooks and used for self-assessment later in the investigation.

In the **Diversity of Life Course**, you ask,

➤ *What is life?*

For a quick write, students write an answer immediately, before instruction occurs. The quick write provides insight into what students think about certain phenomena before you begin instruction. When responding to the question, students should be encouraged to write down their thoughts, even if they don't feel confident in knowing the answer.

Knowing students' preconceptions will help you know what concepts need the most attention during the investigation. Make sure students date their entries for later reference. Read through students' writing and tally the important points to focus on. Quick writes should not be graded.

Planning. After students enter the focus question or complete a quick write in their notebooks, they plan their investigation. (In some investigations, planning is irrelevant to the task at hand.) Planning may be detailed or intuitive, formal or informal, depending on the requirement of the investigation. Plans might include lists (including materials, things to remember), step-by-step procedures, and experimental design. Some FOSS notebook masters guide students through a planning process specific to the task at hand.

Lists. Science notebooks often include lists of things to think about, materials to get, or words to remember. A materials list is a good organizer that helps students anticipate actions they will take. A list of variables to be controlled clarifies the purpose of an experiment. Simple lists of dates for observations or of the people responsible for completing a task may be useful.

Step-by-step procedures. Middle school students need to develop skills for writing sequential procedures. For example, in the **Chemical Interactions Course**, students write a procedure to answer these questions.

- *Is there a limit to the amount of substance that will dissolve in a certain amount of liquid?*
- *If so, is the amount that will dissolve the same for all substances?*

Students need to recall what they know about the materials, develop a procedure for accurately measuring the amount of a substance that is added to the water, and agree on a definition of "saturated." To check a procedure for errors or omissions, students can trade notebooks and attempt to follow another student's instructions to complete the task.

Experimental design. Some work with materials requires a structured experimental plan. In the **Planetary Science Course**, students pursue this focus question.

- *Are Moon craters the result of volcanoes or impacts?*

Students plan an experiment to determine what affects the size and shape of craters on the Moon. They use information they gathered during the open exploration of craters made in flour to develop a strategy for evaluating the effect of changing the variable of a projectile's height or mass. Each lab group agrees on which variable they will change and then designs a sound experimental procedure that they can refer to during the active investigation.

Data Acquisition and Organization

Because observation is the starting point for answering the focus question, data records should be

- clearly related to the focus question;
- accurate and precise;
- organized for efficient reference.

Data handling can have two subcomponents: data acquisition and data display. Data acquisition is making and recording observations (measurements). The data record can be composed of words, phrases, numbers, and drawings. Data display reorganizes the data in a logical way to facilitate thinking. The display can be a graph, chart, calendar, or other graphic organizer.

Early in a student's experience with notebooks, the record may be disorganized and incomplete, and the display may need guidance. The FOSS notebook masters are designed to help students with data collection and organization. You may initially introduce conventional data-display methods, such as those found in the FOSS notebook masters, but soon students will need opportunities to independently select appropriate data displays. As students become more familiar with collecting and organizing data, you might have them create their own records. With practice, students will become skilled at determining what form of recording to use in various situations, and how best to display the data for analysis.

Narratives. For most students, the most intuitive approach to recording data is narrative—using words, sentence fragments, and numbers in a more or less sequential manner. As students make a new observation, they record it below the previous entry, followed by the next observation, and so on. Some observations, such as a record of weather changes in the **Weather and Water Course** or the interactions of organisms in miniecosystems in the **Populations and Ecosystems Course**, are appropriately recorded in narrative form.

Drawings. A picture is worth a thousand words, and a labeled picture is even more useful. When students use a microscope to discover cells in the *Elodea* leaf and observe and draw structures of microorganisms in the **Diversity of Life Course**, a labeled illustration is the most efficient way to record data.

Charts and tables. An efficient way to record many kinds of data is a chart or table. How do you introduce this skill into the shared knowledge of the classroom? One way is to call for attention during an investigation and demonstrate how to perform the operation. Or you can let students record the data as they like, and observe their methods. There may be one or more groups that invent an appropriate table. During processing time, ask this group to share its method with the class. If no group has spontaneously produced an effective table, you might challenge the class to come up with an easier way to display the data, and turn the skill-development introduction into a problem-solving session.

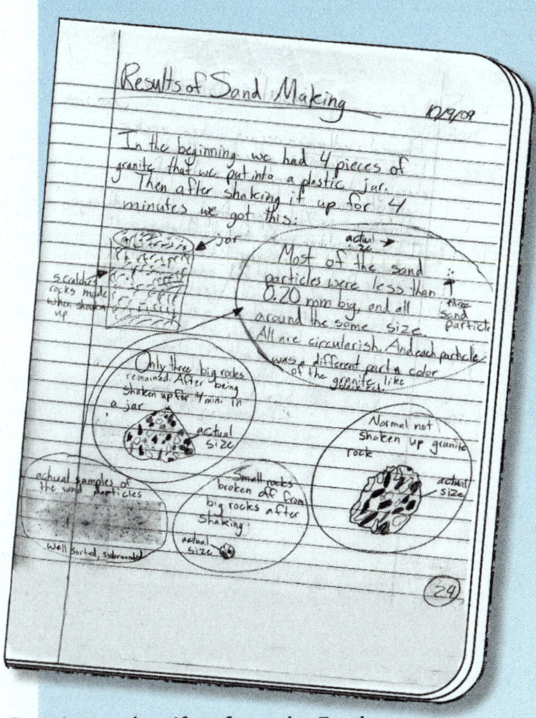

Drawing and artifact from the Earth History Course

With experience, students will recognize when a table or chart is appropriate for recording data. When students make similar observations on a series of objects, such as rock samples in the **Earth History Course**, a table with columns is an efficient way to organize observations for easy comparison.

Artifacts. Occasionally an investigation will produce two-dimensional artifacts that students can tape or glue directly into a science notebook. The mounted flower parts in the **Diversity of Life Course** and the sand samples card from the **Earth History Course** can become a permanent part of the record of learning.

Graphs and graphic tools. Reorganizing data into logical, easy-to-use graphic tools is typically necessary for data analysis. Graphs allow easy comparison (bar graph), quick statistical analysis of frequency data (histogram or line plot), and visual confirmation of a relationship between variables (two-coordinate graph). The **Variables and Design Course** offers many opportunities for students to collect data and organize the data into graphs. Students collect data from air trolleys traveling at different speeds, graph the data, and use the resulting graphs to understand how slope of a motion graph can indicate speed. Other graphic tools, such as Venn diagrams, pie charts, and concept maps, help students make connections.

Making Sense of Data

After collecting and organizing data, the student's next task is to answer the focus question. Students can generate an explanation as an unassisted narrative, but in many instances you might need to use supports such as the FOSS notebook masters to guide the development of a coherent and complete response to the question. Several other support structures for sense making are described below.

Development of vocabulary. Vocabulary is better introduced after students have experienced the new word(s) in context. This sequence provides a cognitive basis for students to connect accurate and precise language to their real-life experiences. Lists of new vocabulary words in the index reinforce new words and organize them for easy reference.

Data analysis. Interpreting data requires the ability to look for patterns, trends, outliers, and potential causes. Students should be encouraged to develop a habit of looking for patterns and relationships within the data collected. Frequently, this is accomplished by creating a graph with numerical data. In the **Populations and Ecosystems Course**, students review field data acquired by ecologists at Mono Lake to determine how biotic and abiotic factors affect the populations of organisms found in the lake.

Graphic organizers. Students can benefit from organizers that help them look at similarities and differences. A compare-and-contrast chart can help students make a transition from their collected data and experiences to making and writing comparisons. It is sometimes easier for students to use than a Venn diagram, and is commonly referred to as a box-and-T chart (as popularized in *Writing in Science: How to Scaffold Instruction to Support Learning*, listed in the Bibliography section).

In this strategy, students draw a box at the top of the notebook page and label it "similar" or "same." On the bottom of the notebook page, they draw a *T*. At the top of each wing of the *T*, they label the objects being compared. Students look at their data, use the *T* to identify differences for each item, and use the "similar" box to list all the characteristics that the two objects have in common. For example, a box-and-T chart comparing characteristics of extrusive and intrusive igneous rocks in the **Earth History Course** might look like this.

similar

extrusive	intrusive

Students can use the completed box-and-T chart to begin writing comparisons. It is usually easier for students to complete their chart on a separate piece of paper, so they can fill it in as they refer to their data. They affix the completed chart into their notebooks after they have made their comparisons.

Claims and evidence. A claim is an assertion about how the natural world works. Claims should always be supported by evidence—statements that are directly correlated with data. The evidence should refer to specific observations, relationships that are displayed in graphs, tables of data that show trends or patterns, dates, measurements, and so on. A claims-and-evidence construction is a sophisticated, rich display of student learning and thinking. It also shows how the data students collected is directly connected to what they learned.

Frames and prompts. One way to get students to organize their thinking is by providing sentence frames for them to complete.

- *I used to think ______, but now I think ______.*
- *The most important thing to remember about Moon phases is ______.*
- *One new thing I learned about adaptation is ______.*

Prompts also direct students to the content they should be thinking about, but provide more latitude for generating responses. For students who are learning English or who struggle with writing, assistive structures like sentence frames can help them communicate their thinking while they learn the nuances of science writing. The prompts used most often in the FOSS notebook masters take the form of questions for students to answer. In the **Weather and Water Course**, students answer the quick-write question

➤ *What causes seasons?*

After modeling an Earth/Sun system and reviewing solar angle and solar concentration, students revisit their quick write to revise and expand on their original explanations.

- *I used to think seasons were caused by ______, but now I know ______.*

Careful prompts scaffold students by helping them communicate their thinking but do not do the thinking for them. As students progress in communication ability, you might provide frames less frequently.

Conclusions and predictions. At the end of an investigation (major conceptual sequence), it might be appropriate for students to write a summary to succinctly communicate what they have learned. This is where students can make predictions based on their understanding of a principle or relationship. For instance, after completing the investigation of condensation and dew point in the **Weather and Water Course**, a student might predict the altitude at which clouds would form, based on weather-balloon data. Or, after examining ecosystem interactions between biotic and abiotic factors in the **Populations and Ecosystems Course**, students will predict how various human interactions could affect the ecosystem. The conclusion or prediction will frequently indicate the degree to which a student can apply new knowledge to real-world situations. A prediction can also be the springboard for further inquiry.

Generating new questions. Does the investigation connect to a student's personal interests? Does the outcome suggest a question or pique a student's curiosity? The science classroom is most exciting when students are generating their own questions for further investigation based on class or personal experiences. The notebook is an excellent place to capture students' musings and record thoughts that might otherwise be lost.

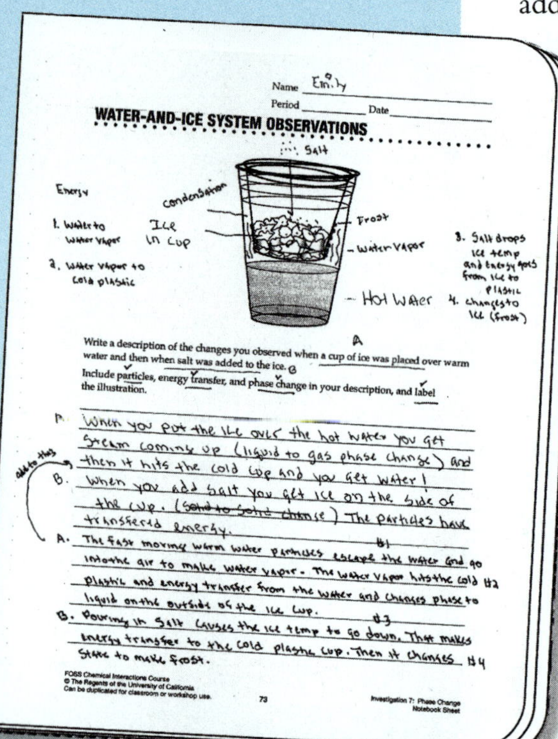

Name Emily
Period ________ Date ________

WATER-AND-ICE SYSTEM OBSERVATIONS

Salt

Energy

1. Water to water vapor

2. Water vapor to cold plastic

condensation

Ice in cup

Frost

Water vapor

Hot water

3. Salt drops ice temp and energy goes from ice to plastic

4. changes to ice (frost)

Write a description of the changes you observed when a cup of ice was placed over warm water and then when salt was added to the ice. Include particles, energy transfer, and phase change in your description, and label the illustration.

A. When you put the ice over the hot water you get steam coming up (liquid to gas phase change) and then it hits the cold cup and you get water!

B. When you add salt you get ice on the side of the cup. (~~solid to solid~~ change) The particles have transfered energy.

A. The fast moving warm water particles escape the water and go into the air to make water vapor. #1 The water vapor hits the cold plastic and energy transfer from the water and changes phase to liquid on the outside of the ice cup. #2

B. Pouring in salt causes the ice temp to go down. #3 That makes energy transfer to the cold plastic cup. Then it changes state to make frost. #4

FOSS Chemical Interactions Course
© The Regents of the University of California
Can be duplicated for classroom or workshop use.

73

Investigation 7: Phase Change Notebook Sheet

A student's revised work for the Chemical Interactions Course

Next-Step Strategies

The goal of the FOSS curriculum is for students to develop accurate, durable knowledge of the science content under investigation. Students' initial conceptions are frequently incomplete or confused, requiring additional thought to become fully functional. The science notebook is a useful place to guide reflection and revision. Typically students commit their understanding in writing and reflect in three locations.

- Explanatory narratives in notebooks
- Response sheets incorporated into the notebook
- Written work on I-Checks

These three categories of written work provide information about student learning for you *and* a record of thinking for students that they can reflect on and revise. Scientists constantly refine and clarify their ideas about how the natural world works. They read scientific articles, consult with other scientists, and attend conferences. They incorporate new information into their thinking about the subject they are researching. This reflective process can result in deeper understanding or a complete revision of thinking.

After completing one of the expositions of knowledge—a written conclusion, response sheet, or benchmark assessment—students should receive additional instruction or information via a next-step strategy. They will use this information later to complete self-assessment by reviewing their original written work, making judgments about its accuracy and completeness, and writing a revised explanation. You can use any of a number of techniques for providing the additional information to students.

- Group compare-and-share discussion
- Think/pair/share reading
- Whole-class critique of an explanation by an anonymous student
- Identifying key points for a class list
- Whole-class discussion of a presentation by one student

After one of the information-generating processes, students compare the "best answer" to their own answer and rework their explanations if they can no longer defend their original thinking. The revised statement of the science content can take one of several forms.

Students might literally revise the original writing, crossing out extraneous or incorrect bits, inserting new or improved information, and completing the passage. At other times, students might reflect on their original work and, after drawing and dating a line of learning (see below), might redraft their explanation from scratch, producing their best explanation of the concept.

During these self-assessment processes, students have to think actively about every aspect of their understanding of the concept and organize their thoughts into a coherent, logical narrative. The learning that takes place during this process is powerful. The relationships between the several elements of the concept become unified and clarified.

The notebook is the best tool for students when preparing for benchmark assessment, such as an I-Check or posttest. Students don't necessarily have the study skills needed to prepare on their own, but using teacher-guided tasks such as key points and traffic lights will turn the preparation process into a valuable exercise. These same strategies can be used after a benchmark assessment when you identify further areas of confusion or misconceptions you want to address with students. Here are four helpful next-step, or self-assessment, strategies.

Name ____________
Period ______ Date ____________

RESPONSE SHEET—WHERE AM I?

Billy was watching his favorite TV show when the question came up, "Where do you live?" Billy didn't know, so he asked his older sister. She replied, "You live in Birmingham."

Later Billy asked his mom the same question, to which she replied, "Honey, you live in the United States."

Billy was a little confused. He thought, "How can I live in two places at the same time?"

Who was right, Billy's sister? Billy's mom? Explain your answer.

Billy, They are both right. Biminghom is in the united states. its Just a different kind of way of saying it

frame of refrence

Silly billy, They both just said a different fame of reference. Your sister said that you live in burmingh. and your mom said you live in the us. its just different frame of refrence

A line of learning used with the Planetary Science Course

Line of learning. One technique many teachers find useful in the reflective process is the line of learning. After students have conducted an investigation and entered their initial explanations, they draw and date a line under their original work. As students share ideas and refine their thinking during class discussion, additional experimentation, reading, and teacher feedback, encourage them to make new entries under the line of learning, adding to or revising their original thinking. If the concept is elusive or complex, a second line of learning, followed by more processing and revising, may be appropriate.

The line of learning is a reminder to students that learning is an ongoing process with imperfect products. It points out places in that process where a student made a stride toward full understanding. And the psychological security provided by the line of learning reminds students that they can always draw another line of learning and revise their thinking again. The ability to look back in the science notebook and see concrete evidence of learning gives students confidence and helps them become critical observers of their own learning.

Traffic lights. In the traffic-lights strategy, students use color to self-assess and indicate how well they understand a concept that they are learning. Green means that the student feels that he or she has a good understanding of the concept. Yellow means that the student is still a bit unsure about his or her understanding. Red means that the student needs help; he or she has little or no understanding of the concept. Students can use colored pencils, markers, colored dots, or colored cards to indicate their understanding. They can mark their own work and then indicate their level of understanding by a show of hands or by holding up colored cards. This strategy gives students practice in self-assessment and helps you monitor students' current understanding. You should follow up by looking at student work to ensure that they actually do understand the content that they marked with green.

Three C's. Another approach to revision is to apply the three C's—confirm, correct, complete—to the original work. Students indicate ideas that were correct with a number or a color, code statements needing correction with a second number or color, and assign a third number or color to give additional information that completes the entry.

Key points. Students do not necessarily connect the investigative experience with the key concepts and processes taught in the lesson. It is essential to give students an opportunity to reflect on their experiences and find meaning in those experiences. They should be challenged to use their experiences and data to either confirm or reject their current understanding of the natural world. As students form supportable ideas about a concept, those ideas should be noted as key points, posted in the room, and written in their notebooks. New evidence, to support or clarify an idea, can be added to the chart as the course progresses. If an idea doesn't hold up under further investigation, a line can be drawn through the key point to indicate a change in thinking. A key-points activity is embedded near the end of each investigation to help students organize their thinking and prepare for benchmark assessment.

USING NOTEBOOKS TO IMPROVE STUDENT LEARNING

Notebook entries should not be graded. Research has shown that more learning occurs when students get only comments on written work in their notebooks, not grades or a combination of comments and grades.

If your school district requires a certain number of grades each week, select certain work products that you want to grade and have students turn in that work separate from the notebook. After grading, return the piece to students to insert into their notebooks, so that all their work is in one place.

It may be difficult to stop using grades or a rubric for notebook assessment. But providing feedback that moves learning forward, however difficult, has benefits that make it worth the effort. The key to using written feedback for formative assessment is to make feedback timely and specific, and to provide time for students to act on the feedback by revising or correcting work right in their own notebook.

Teacher Feedback

Student written work often exposes weaknesses in understanding—or so it appears. It is important for you to find out if the flaw results from poor understanding of the science or from imprecise communication. You can use the notebook to provide two types of feedback to the student: to ask for clarification or additional information, and to ask probing questions that will help students move forward in their thinking. Respecting the student's space is important, so rather than writing directly in the notebook, attach a self-stick note, which can be removed after the student has taken appropriate action.

The most effective forms of feedback relate to the content of the work. Here are some examples.

- *You wrote that seasons are caused by Earth's tilt. Does Earth's tilt change during its orbit?*
- *What evidence can you use to support your claim that Moon craters are caused by impacts? Hint: Think of our experiments in class.*

Nonspecific feedback, such as stars, pluses, smiley faces, and "good job!", or ambiguous critiques, such as "try again," "put more thought into this," and "not enough," are less effective and should not be used. Feedback that guides students to think about the content of their work and gives suggestions for how to improve are productive instructional strategies.

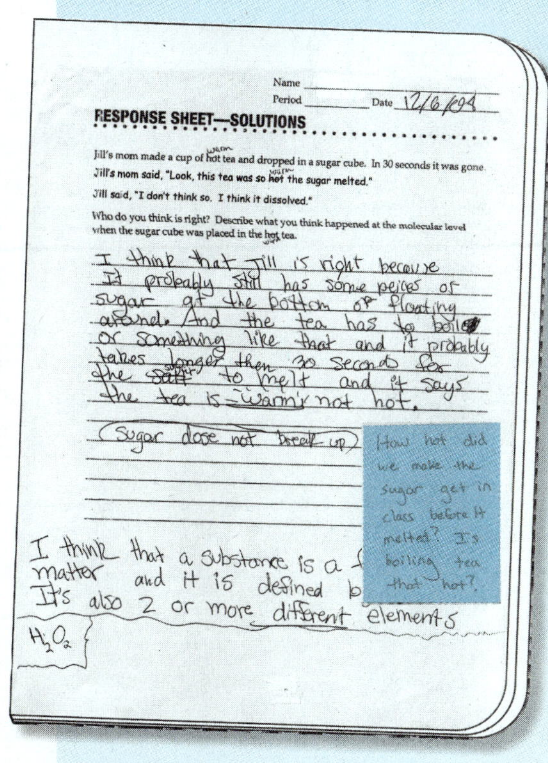

Name
Period Date 12/6/09

RESPONSE SHEET—SOLUTIONS

Jill's mom made a cup of hot warm tea and dropped in a sugar cube. In 30 seconds it was gone.

Jill's mom said, "Look, this tea was so hot warm the sugar melted."

Jill said, "I don't think so. I think it dissolved."

Who do you think is right? Describe what you think happened at the molecular level when the sugar cube was placed in the hot warm tea.

I think that Jill is right because It probably still has some peices of sugar at the bottom or floating around. And the tea has to boil or something like that and it probably takes longer then 30 seconds for the salt sugar to melt and it says the tea is "warm" not hot.

(Sugar dose not break up)

How hot did we make the sugar get in class before it melted? Is boiling tea that hot?

I think that a substance is a ... matter and it is defined b... It's also 2 or more different elements

H_2O_2

Feedback given during the Chemical Interactions Course

Here are some appropriate generic feedback questions to write or use verbally while you circulate in the class.

- *What vocabulary have you learned that will help you describe ______?*
- *Can you include an example from class to support your ideas?*
- *Include more detail about ______.*
- *Check your data to make sure this is accurate.*
- *What do you mean by ______?*
- *When you record your data, what unit should you use?*

When students return to their notebooks and respond to the feedback, you will have additional information to help you discriminate between learning and communication difficulties. Another critical component of teacher feedback is providing comments to students in a timely manner, so that they can review their work before engaging in benchmark assessment or moving on to other big ideas in the course.

In middle school, you face the challenge of having a large number of students. This may mean collecting a portion of students' notebooks on alternate days. Set a specific focus for your feedback, such as a data table or conclusion, so you aren't trying to look at everything every time.

To help students improve their writing, you might have individuals share notebook entries aloud in their collaborative groups, followed by feedback from a partner or the group. This valuable tool must be very structured to create a safe environment, including ground rules about acceptable feedback and comments.

A good way to develop these skills is to model constructive feedback with the class, using a student-work sample from a notebook. Use a sample from a previous year with the name and any identifying characteristics removed. Project it for the class to practice giving feedback.

Formative Assessment

With students recording more of their thinking in an organized notebook, you have a tool to better understand the progress of students and any misconceptions that are typically not revealed until the benchmark assessment. One way to monitor student progress is during class while they are responding to a prompt. Circulate from group to group, and read notebook entries over students' shoulders. This is a good time to have short conversations with individuals or small groups to gain information about the level of student understanding. Take care to respect the privacy of students who are not comfortable sharing their work during the writing process.

If you want to look at work that is already completed in the notebook, ask students to open their notebooks to the page that you want to review and put them in a designated location. Or consider having students complete the work on a separate piece of paper or an index card. Students can leave a blank page in their notebooks, or label it with a header as a placeholder, until they get the work back and tape it or glue it in place. This makes looking at student work much easier, and the record of learning that the student is creating in the notebook remains intact.

When time is limited, you might select a sample of students from each class, alternating the sample group each time, to get a representative sample of student thinking. This is particularly useful following a quick write.

Once you have some information about student thinking, you can make teaching decisions about moving ahead to a benchmark assessment, going back to a previous concept, or spending more time making sense of a concept. Benchmark assessments can also be used as formative assessment. You might choose to administer an I-Check, score the assessment to find problem areas, and then revisit critical concepts before moving on to the next investigation. Students can use reflection and self-assessment techniques to revisit and build on their original exam responses.

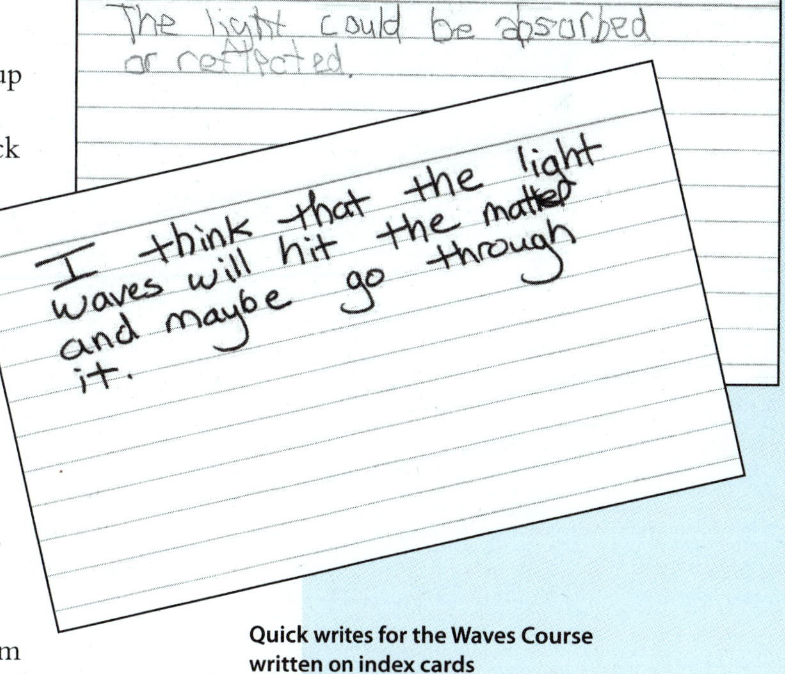

Quick writes for the Waves Course written on index cards

DERIVATIVE PRODUCTS

On occasion, you might ask students to produce science projects of various kinds: summary reports, detailed explanations, end-of-course projects, oral reports, or posters. Students should use their notebooks as a reference when developing their reports. You could ask them to make a checklist of science concepts and pieces of evidence, with specific page references, extracted from their notebooks. They can then use this checklist to ensure that all important points have been included in the derivative work.

The process of developing a project has feedback benefits, too. While students are developing projects using their notebooks, they have the opportunity to self-monitor the organization and content of the notebook. This offers valuable feedback on locating and extracting useful information. You might want to discuss possible changes students would make next time they start a new science notebook.

Homework is another form of derivative product, as it is an extension of the experimentation started in class. Carefully selected homework assignments enhance students' science learning. Homework suggestions and/or extension activities are included at the end of each investigation. For example, in the **Heredity and Adaptation Course**, after using an online activity in class to predict genetic variation, students are asked to complete a follow-up online simulation as homework. In the **Electromagnetic Force Course**, after students test properties of magnets in class, they are asked to look for examples of magnets in household objects outside of the classroom.

Science-Centered Language Development in Middle School

Reading and writing are inextricably linked to the very nature and fabric of science, and, by extension, to learning science.

Stephen P. Norris and Linda M. Phillips, "How Literacy in Its Fundamental Sense Is Central to Scientific Literacy"

Contents

INTRODUCTION

In this chapter, we explore the ways reading, writing, speaking, and listening are interwoven in effective science instruction at the secondary level. To engage fully in the enterprise of science and engineering, students must record and communicate observations and explanations, and read about and discuss the discoveries and ideas of others. This becomes increasingly challenging at the secondary level. Texts become more complex; writing requires fluency of academic language, including domain-specific vocabulary. Here we identify strategies that support sense making. The active investigations, student science notebooks, *FOSS Science Resources* readings, multimedia, and formative assessments provide rich contexts in which students develop and exercise thinking processes and communication skills. Students develop scientific literacy through experiences with the natural world around them in real and authentic ways and use language to inquire, process information, and communicate their thinking about the objects, organisms, and phenomena they are studying. We refer to the acquisition and building of language skills necessary for scientific literacy as science-centered language development.

Language plays two crucial roles in science learning: (1) it facilitates the communication of conceptual and procedural knowledge, questions, and propositions (external; public), and (2) it mediates thinking, a process necessary for understanding (internal; private). These are also the ways scientists use language: to communicate with one another about their inquiries, procedures, and understandings; to transform their observations into ideas; and to create meaning and new ideas from their work and the work of others. For students, language development is intimately involved in their learning about the natural world. Active-learning science provides a real and engaging context for developing literacy; language-arts skills and strategies support conceptual development and scientific and engineering practices. For example, the skills and strategies used for reading comprehension, writing expository text, and oral discourse are applied when students are recording their observations, making sense of science content, and communicating their ideas. Students' use of language improves when they discuss, write, and read about the concepts explored in each investigation.

We begin our exploration of science and language by focusing on language functions and how specific language functions are used in science to facilitate information acquisition and processing (thinking). Then we address issues related to the specific language domains—speaking and listening, writing, and reading. Each section addresses

- how skills in that domain are developed and exercised in FOSS science investigations;
- literacy strategies that are integrated purposefully into the FOSS investigations; and
- suggestions for additional literacy strategies that both enhance student learning in science and develop or exercise academic-language skills.

Following the domain discussions is a section on science-vocabulary development, with scaffolding strategies for supporting all learners. The last section covers language-development strategies specifically for English learners.

NOTE

The term *English learners* refers to students who are learning to understand English. This includes students who speak English as a second language and native English speakers who need additional support to use language effectively.

THE ROLE OF LANGUAGE IN SCIENTIFIC AND ENGINEERING PRACTICES

Language functions are the purpose for which speech or writing is used and involve both vocabulary and grammatical structure. Understanding and using language functions appropriately is important in effective communication. Students use numerous language functions in all disciplines to mediate communication and facilitate thinking (e.g., they plan, compare, discuss, apply, design, draw, and provide evidence).

In science, language functions facilitate scientific and engineering practices. For example, when students are *collecting data*, they are using language functions to identify, label, enumerate, compare, estimate, and measure. When students are *constructing explanations*, they are using language functions to analyze, communicate, discuss, evaluate, and justify.

A Framework for K–12 Science Education (National Research Council 2012) states that "Students cannot comprehend scientific practices, nor fully appreciate the nature of scientific knowledge itself, without directly experiencing the practices for themselves." Each of these scientific and engineering practices uses multiple language functions. Often, these language functions are part of an internal dialogue weighing the merits of various explanations—what we call thinking. The more language functions with which we are facile, the more effective and creative our thinking can be.

The scientific and engineering practices are listed below, along with a sample of the language functions that are exercised when students are effectively engaged in that practice. (Practices are bold; language functions are italic.)

Asking questions and defining problems

- *Ask* questions about objects, organisms, systems, and events in the natural and human-made world (science).
- *Ask* questions to *define* and *clarify* a problem, *determine criteria* for solutions, and *identify* constraints (engineering).

Planning and carrying out investigations

- *Plan* and conduct investigations in the laboratory and in the field to gather appropriate data (*describe* procedures, *determine* observations to *record*, *decide* which variables to control) or to gather data essential for *specifying* and *testing* engineering designs.

Examples of Language Functions

Analyze
Apply
Ask
Clarify
Classify
Communicate
Compare
Conclude
Construct
Critique
Describe
Design
Develop
Discuss
Distinguish
Draw
Enumerate
Estimate
Evaluate
Experiment
Explain
Formulate
Generalize
Group
Identify
Infer
Interpret
Justify
Label
List
Make a claim
Measure
Model
Observe
Organize
Plan
Predict
Provide evidence
Reason
Record
Represent
Revise
Sequence
Solve
Sort
Strategize
Summarize
Synthesize

Analyzing and interpreting data

- Use a range of tools (numbers, words, tables, graphs, images, diagrams, equations) to *organize* observations (data) in order to *identify* significant features and patterns.

Developing and using models

- Use models to help *develop explanations, make predictions*, and *analyze* existing systems, and *recognize* strengths and limitations of the models.

Using mathematics and computational thinking

- Use mathematics and computation to *represent* physical variables and their relationships.

Constructing explanations and designing solutions

- *Construct* logical explanations of phenomena, or *propose solutions* that incorporate current understanding or a model that represents it and is consistent with the available evidence.

Engaging in argument from evidence

- *Defend* explanations, *formulate evidence* based on data, *examine* one's own understanding in light of evidence offered by others, and collaborate with peers in searching for explanations.

Obtaining, evaluating, and communicating information

- *Communicate* ideas and the results of inquiry—orally and in writing—with tables, diagrams, graphs, and equations and in *discussion* with peers.

Research supports the claim that when students are intentionally using language functions in thinking about and communicating in science, they improve not only science content knowledge, but also language-arts and mathematics skills (Ostlund, 1998; Lieberman and Hoody, 1998). Language functions play a central role in science as a key cognitive tool for developing higher-order thinking and problem-solving abilities that, in turn, support academic literacy in all subject areas.

Here is an example of how an experienced teacher can provide an opportunity for students to exercise language functions in FOSS. In the **Planetary Science Course**, one piece of content we expect students to have acquired by the end of the course is

- The lower the angle at which light strikes a surface, the lower the density of the light energy.

The scientific practices the teacher wants the class to focus on are *developing and using models* and *constructing explanations*.

The language functions students will exercise while engaging in these scientific practices include *comparing, modeling, analyzing*, and *explaining*. The teacher understands that these language functions are appropriate to the purpose of the science investigation and support the *Common Core State Standards for English Language Arts and Literacy in Science* (CCSS), in which grades 6–8 students will "write arguments focused on discipline-specific content . . . support claim[s] with logical reasoning and relevant, accurate data and evidence that demonstrate an understanding of the topic" (National Governors Association Center for Best Practices, Council of Chief State School Officers, 2010).

- Students will *compare* the area covered by the same beam of light (from a flashlight) at different angles to *explain* the relationship between the angle and density of light energy.

The teacher can support the use of language functions by providing structures such as sentence frames.

- As ______, then ______.

  As the angle increases, then the light beam becomes smaller and more circular.

- When I changed ______, then ______.

  When I changed the angle of the light beam, then the concentration of light hitting the floor changed.

- The greater/smaller ______, the ______.

  The smaller the angle of the light beam, the more the light beam spread out.

- I think ______, because ______.

  I think the smaller spot of light receives more energy than the larger spot because the light concentration is greatest when light shines directly down on a surface and there is no beam spreading.

CCSS NOTE

This example supports CCSS.ELA-Literacy.WHST.6–8.1.b.

SPEAKING AND LISTENING DOMAIN

The FOSS investigations are designed to engage students in productive oral discourse. Talking requires students to process and organize what they are learning. Listening to and evaluating peers' ideas calls on students to apply their knowledge and to sharpen their reasoning skills. Guiding students in instructive discussions is critical to the development of conceptual understanding of the science content and the ability to think and reason scientifically. It also addresses a key middle school CCSS Speaking and Listening anchor standard that students "engage effectively in a range of collaborative discussions (one-on-one, in groups, and teacher-led) with diverse partners on [grade-level] topics, texts, and issues, building on others' ideas and expressing their own clearly."

CCSS NOTE
This example supports CCSS.ELA-Literacy.SL.6.1, CCSS.ELA-Literacy.SL.7.1, and CCSS.ELA-Literacy.SL.8.1.

FOSS investigations start with a discussion—a review to activate prior knowledge, presentation of a focus question, or a challenge to motivate and engage active thinking. During the active investigation, students talk with one another in small groups, share their observations and discoveries, point out connections, ask questions, and start to build explanations. The discussion icon in the sidebar of the *Investigations Guide* indicates when small-group discussions should take place.

Throughout the activity, the *Investigations Guide* indicates where it is appropriate to pause for whole-class discussions to guide conceptual understanding. The *Investigations Guide* provides you with discussion questions to help stimulate student thinking and support sense making. At times, it may be beneficial to use sentence frames or standard prompts to scaffold the use of effective language functions and structures. Allowing students a few minutes to write in their notebooks prior to sharing their answers also helps those who need more time to process and organize their thoughts.

NOTE
Additional notebook strategies can be found in the Science Notebooks in Middle School chapter in *Teacher Resources* and online at www.FOSSweb.com.

On the following pages are some suggestions for providing structure to those discussions and for scaffolding productive discourse when needed. Using the protocols that follow will ensure inclusion of all students in discussions.

Partner and Small-Group Discussion Protocols

Whenever possible, give students time to talk with a partner or in a small group before conducting a whole-class discussion. This provides all students with a chance to formulate their thinking, express their ideas, practice using the appropriate science vocabulary, and receive input from peers. Listening to others communicate different ways of thinking about the same information from a variety of perspectives helps students negotiate the difficult path of sense making for themselves.

Dyads. Students pair up and take turns either answering a question or expressing an idea. Each student has 1 minute to talk while the other student listens. While student A is talking, student B practices attentive listening. Student B makes eye contact with student A, but cannot respond verbally. After 1 minute, the roles reverse.

Here's an example from the **Chemical Interactions Course**. After reviewing the results of eight reactions recorded in their notebooks, you ask students to pair up and take turns sharing which two substances they think constitute the mystery mixture and their reasons for selecting those two. The language objective is for students to compare their test results and make inferences based on their observations and what they know about chemical reactions (orally and in writing). These sentence frames can be written on the board to scaffold student thinking and conversation.

- I think the two substances in the mystery mixture are ______ and ______.
- My evidence is ______.

Partner parade. Students form two lines facing each other. Present a question, an idea, an object, or an image as a prompt for students to discuss. Give them 1 minute to greet the person in front of them and discuss the prompt. After 1 minute, call time. Have the first student in one of the lines move to the end of the line, and have the rest of the students in that line shift one step sideways so that everyone has a new partner. (Students in the other line do not move.) Give students a new prompt to discuss for 1 minute with their new partners. This can also be done by having students form two concentric circles. After each prompt, the inner circle rotates.

For example, when students are just beginning the **Earth History Course** investigation on igneous rock, you may want to assess prior knowledge about Earth's layers. Give each student a picture from an assortment of related images such as volcanoes, magma, a diagram of Earth's layers, crystals, and so forth, and have students line up facing

Partner and Small-Group Discussion Protocols

- *Dyads*
- *Partner parade*
- *Put in your two cents*

each other in two lines or in concentric circles. For the first round, ask, "What do you observe in the image on your card?" For the second round, ask, "What can you infer from the image?" For the third round, ask, "What questions do you have about the image?" The language objective is for students to describe their observations, infer how the landform formed, and reflect upon and relate any experiences they may have had with a similar landform. These sentence frames can be used to scaffold student discussion.

- I observe ______, ______, and ______.
- I think this shows ______ because ______.
- I wonder ______.

Put in your two cents. For small-group discussions, give each student two pennies or similar objects to use as talking tokens. Each student takes a turn putting a penny in the center of the table and sharing his or her idea. Once all have shared, each student takes a turn putting in the other penny and responding to what others in the group have said. For example,

- I agree (or don't agree) with ______ because ______.

Here's an example from the **Diversity of Life Course**. In their notebooks, students have recorded the amount of water lost from their vials containing celery with and without leaves. They discover a discrepancy in the amount of water lost and the mass of the celery. Where did the water go? Students are struggling to form an explanation. The language objective is for students to compare their results and infer that there is a relationship between the amount of water lost and the number of leaves the celery has. You give each student two pennies, and in groups of four, they take turns putting in their two cents. For the first round, each student answers the question "Where did the water go?" They use the frame

- I think the water ______.
- My evidence is ______.

On the second round, each student states whether he or she agrees or disagrees with someone else in the group and why, using the sentence frame.

Whole-Class Discussion Supports

The whole-class discussion is a critical part of sense making. After students have had the active learning experience and have talked with their peers in pairs and/or small groups, sharing their observations with the whole class sets the stage for developing conventional explanatory models. Discrepant events, differing results, and other surprises are discussed, analyzed, and resolved. It is important that students realize that science is a process of finding out about the world around them. This is done through asking questions, testing ideas, forming explanations, and subjecting those explanations to logical scrutiny, that is, argumentation. Leading students through productive discussion helps them connect their observations and the abstract symbols (words) that represent and explain those observations. Whole-class discussion also provides an opportunity for you to interject an accurate and precise verbal summary as a model of the kind of thinking you are seeking. Facilitating effective whole-class discussions takes skill, practice, a shared set of norms, and patience. In the long run, students will have a better grasp of the content and will improve their ability to think independently and communicate effectively.

Norms should be established so that students know what is expected during science discussions.

- Science content and practices are the focus.
- Everyone participates (speaking and listening).
- Ideas and experiences are shared, accepted, and valued. Everyone is respectful of one another.
- Claims are supported by evidence.
- Challenges (debate and argument) are part of the quest for complete understanding.

A variety of whole-class discussion techniques can be introduced and practiced during science instruction that address the CCSS Speaking and Listening standards for students to "present claims and findings [e.g., argument, narrative, informative, summary presentations], emphasizing salient points in a focused, coherent manner with relevant evidence, sound valid reasoning, and well-chosen details; use appropriate eye contact, adequate volume, and clear pronunciation."

For example, during science talk, students are reminded to practice attentive listening, stay focused on the speaker, ask questions, and respond appropriately. In addition, in order for students to develop and practice their reasoning skills, they need to know the language forms

Whole-Class Discussion Supports

- *Sentence frames*
- *Guiding questions*

TEACHING NOTE

Let students know that scientists change their minds based on new evidence. It is expected that students will revise their thinking, based on evidence presented in discussions.

CCSS NOTE

This example supports CCSS.ELA-Literacy.SL.6.4, CCSS.ELA-Literacy.SL.7.4, and CCSS.ELA-Literacy.SL.8.4, (grade 8 quoted here).

TEACHING NOTE

Encourage "science talk." Allow time for students to engage in discussions that build on other students' observations and reasoning. After an investigation, use a teacher- or student-generated question, and either just listen or facilitate the interaction with questions to encourage expression of ideas among students.

and structures and the behaviors used in evidence-based debate and argument, such as using data to support claims, disagreeing respectfully, and asking probing questions (Winokur and Worth, 2006).

Explicitly model the language structures appropriate for active discussions, and encourage students to use them when responding to guiding questions and during science talks.

Sentence frames. These samples can be posted as a scaffold as students develop their reasoning and oral participation skills.

- I think ______, because ______.
- I predict ______, because ______.
- I claim ______; my evidence is ______.
- I agree with ______ that ______.
- My idea is similar/related to ______'s idea.
- I learned/discovered/heard that ______.
- <Name> explained ______ to me.
- <Name> shared ______ with me.
- We decided/agreed that ______.
- Our group sees it differently, because ______.
- We have different observations/results. Some of us found that ______. One group member thinks that ______.
- We had a different approach/idea/solution/answer: ______.

Guiding questions. The Investigations Guide provides questions to help concentrate student thinking on the concepts introduced in the investigation. Guiding questions should be used during the whole-class discussion to facilitate sense making. Here are some other open-ended questions that help guide student thinking and promote discussion.

- What did you notice when ______?
- What do you think will happen if ______?
- How might you explain ______? What is your evidence?
- What connections can you make between ______ and ______?

Whole-Class Discussion Protocols

The following tried-and-true participation protocols can be used to enhance whole-class discussions. The purpose of these protocols is to increase meaningful participation by giving all students access to the discussion, allowing students time to think (process), and providing a context for motivation and engagement.

Think-pair-share. When asking for a response to a question posed to the class, give students a minute to think silently. Then, have students pair up with a partner to exchange thoughts before you call on a student to share his or her ideas with the whole class.

Pick a stick. Write each student's name on a craft stick, and keep the sticks handy in a cup at the front of the room. When asking for responses, randomly pick a stick, and call on that student to start the discussion. Continue to select sticks as you continue the discussion. Your name can also be on a stick in the cup. To keep students on their toes, put the selected sticks into a smaller cup hidden inside the larger cup out of view of students. That way students think they may be called again.

Whip around. Each student takes a quick turn sharing a thought or reaction. Questions are phrased to elicit quick responses that can be expressed in one to five words (e.g., "Give an example of a stored-energy source." "What does the word *heat* make you think of?").

Commit and toss. Have students write a response to a question or prompt on a loose piece of paper (Keeley, 2008). Next, tell everyone to crumple up the paper into a ball and toss it to another student. Continue tossing for a few minutes, and then call for students to stop, grab a ball, and read the response silently. Responses can then be shared with partners, small groups, or the whole class. This activity allows students to answer anonymously, so they may be willing to share their thinking more openly.

Group posters. Have small groups design and graphically record their investigation data and conclusions on a quickly generated poster to share with the whole class.

Whole-Class Discussion Protocols
- *Think-pair-share*
- *Pick a stick*
- *Whip around*
- *Commit and toss*
- *Group posters*

Cup within a cup pick-a-stick container

WRITING DOMAIN

Information processing is enhanced when students engage in informal writing. When allowed to write expressively without fear of being scorned for incorrect spelling or grammar, students are more apt to organize and express their thoughts in different ways that support their own sense making. Writing in science promotes use of science and engineering practices, thereby developing a deeper engagement with the science content. This type of informal writing also provides a springboard for more formal derivative science writing (Keys, 1999).

Science Notebooks

The science notebook is an effective tool for enhancing learning in science and exercising various forms of writing. Notebooks provide opportunities both for expressive writing (students craft explanatory narratives that make sense of their science experiences) and for practicing informal technical writing (students use organizational structures and writing conventions). Students learn to communicate their thinking in an organized fashion while engaging in the cognitive processes required to develop concepts and build explanations. Having this developmental record of learning also provides an authentic means for assessing students' progress in both scientific thinking and communication skills.

NOTE
For more information about supporting science-notebook development, see the Science Notebooks in Middle School chapter.

Developing Writing for Literacy in Science

Using student science notebooks in science instruction provides opportunities to address the CCSS for Writing in Science. Grades 6–8 students "write routinely over extended time frames (time for research, reflection, and revision) and shorter time frames (a single sitting or a day or two) for a range of tasks, purposes, and audiences." In addition to providing a structure for recording and analyzing data, notebooks serve as a reference tool from which students can draw information in order to produce derivative products, that is, more formal science writing pieces that have a specific purpose and format. CCSS focus on three text types that students should be writing in science: argument, informational/explanatory writing, and narrative writing. These text types are used in science notebooks and can be developed into derivative products such as reports, articles, brochures, poster boards, electronic presentations, letters, and so forth. Following is a description of these three text types and examples that may be used with FOSS investigations to help students build scientific literacy.

CCSS NOTE
This example supports CCSS.ELA-Literacy.W.10.

Engaging in Argument

In science, middle school students make claims in the form of statements or conclusions that answer questions or address problems. CCSS Appendix A describes that for students to use "data in a scientifically acceptable form, students marshal evidence and draw on their understanding of scientific concepts to argue in support of their claims." Applying the literacy skills necessary for this type of writing concurrently supports the development of critical science and engineering practices—most notably, engaging in argument. According to *A Framework for K–12 Science Education*, upon which the Next Generation Science Standards (NGSS) are based, middle school students are expected to construct a convincing argument that supports or refutes claims for explanations about the natural and designed world in these ways.

In FOSS, this type of writing makes students' thinking visible. Both informally in their notebooks and formally on assessments, students use deductive and inductive reasoning to construct and defend their explanations. In this way, students deepen their science understanding and exercise the language functions necessary for higher-level thinking, for example, comparing, synthesizing, evaluating, and justifying. To support students in both oral and written argumentation, use the questions and prompts in the *Investigations Guide* that encourage students to use evidence, models, and theories to support their arguments. In addition, be prepared for those teachable moments that provide the perfect stage for spontaneous scientific debate. Here are some general questions to help students deepen their writing.

- Why do you agree or disagree with ______?
- How would you prove/disprove ______?
- What data did you use to make that conclusion that ______?
- Why was it better that ______?

Here are the ways engaging in written argument are developed in the FOSS investigations and can be extended through formal writing.

Response sheets. The FOSS response sheets give students practice in constructing arguments by providing hypothetical situations where they have to apply what they have learned in order to evaluate a claim. For example, one of the response sheets in the **Planetary Science Course** asks students to respond to three students' explanations for the seasons. Students write a paragraph to each student with the

Engaging in Argument
- *Response sheets*
- *Think questions*
- *I-Checks and surveys/posttests*
- *Persuasive writing*

CCSS NOTE
This example supports CCSS.ELA-Literacy.W.1.

purpose of changing his or her thinking. In order to refute each claim, students must evaluate the validity of the statements and construct arguments based on evidence from the data they've collected during the investigations and logical reasoning that supports their explanation for what causes seasons.

Think questions. Interactive reading in *FOSS Science Resources* is another opportunity for students to engage in written argumentation. Articles include questions that support reading comprehension and extend student thinking about the science content. Asking students to make a claim and provide evidence to support it encourages the use of language functions necessary for higher-level thinking such as evaluating, applying, and justifying. For example, in *FOSS Science Resources: Planetary Science*, students are asked to respond to the following question: Why do you think there are so few craters on Earth and so many on the Moon? After discussion with their peers, students can hone their argumentation skills by writing an argument that answers the question and is supported by the evidence in the *FOSS Science Resources* book as well as data recorded from their experience making model craters.

I-Checks and surveys/posttests. Like the FOSS response sheets, some test items assess students' ability to make a claim and provide evidence to support it. One way is to provide students with data and have them make a claim based on that data and evidence from their prior investigations. Their argument should use logical reasoning to support their ideas. For example, in **Planetary Science**, students are shown images taken from two different planets. They are told that one has a thick atmosphere and the other has no atmosphere. They are asked which image they think came from a planet with an atmosphere and why. Using the images, they can see evidence of craters, and they can draw on their own experiences as well as knowledge acquired through other sources to piece together a logical argument.

Persuasive writing. Formal writing gives students the opportunity to summarize, explain, apply, and evaluate what they have learned in science. It also provides a purpose and audience that motivate students to produce higher-level writing products. The objective of persuasive writing is to convince the reader that a stated interpretation of data is worthwhile and meaningful. In addition to supporting claims with evidence and using logical argument, the writer also uses persuasive techniques such as a call to action. Students can use their informal notebook entries to form the basis of formal persuasive writing in a variety of formats, such as essays, letters, editorials, advertisements, award nominations, informational pamphlets, and petitions. Animal habitats, energy use, weather patterns, landforms, and water sources are just a few science topics that can generate questions and issues for persuasive writing.

Here is a sample of writing frames that can be used to introduce and scaffold persuasive writing (modified from Gibbons, 2002).

Title: ______

The topic of this discussion is ______.

My opinion (position, conclusion) is ______.

There are <number> reasons why I believe this to be true.

First, ______.

Second, ______.

Finally, ______.

On the other hand, some people think ______.

I have also heard people say ______.

However, my claim is that ______ because ______.

Informational/Explanatory Writing
- *Writing frames*
- *Recursive cycle*

Informational/Explanatory Writing

Informational and explanatory writing requires students to examine and convey complex ideas and information clearly and accurately through the effective selection, organization, and analysis of content. In middle school science, this includes writing scientific procedures and experiments. Described in CCSS Appendix A, informational/explanatory writing answers the questions, What type? What are the components? What are the properties, functions, and behaviors? How does it work? What is happening? Why? In FOSS, this type of writing takes place informally in science notebooks, where students are recording their questions, plans, procedures, data, and answers to the focus questions. It also supports sense making as students attempt to convey what they know in response to questions and prompts, using language functions such as identifying, comparing and contrasting, explaining cause-and-effect relationships, and sequencing.

CCSS NOTE
Designing, recording, and following procedures in FOSS courses supports CCSS.ELA-Literacy.RST.6–8.3.

As an extension of the notebook entries, students can apply their content knowledge to publish formal products such as letters, definitions, procedures, newspaper and magazine articles, posters, pamphlets, and research reports. Strategies such as the writing process (plan, draft, edit, revise, and share) and writing frames (modeling and guiding the use of topic sentences, transition and sequencing words, examples, explanations, and conclusions) can be used to scaffold and help students develop proficiency in science writing.

Writing frames. Here are samples of writing frames (modified from Wellington and Osborne, 2001).

CCSS NOTE
This example supports CCSS.ELA-Literacy.W.2.

Description

Title: ______

(Identify) The part of the ______ I am describing is the ______.

(Describe) It consists of ______.

(Explain) The function of these parts is ______.

(Example) This drawing shows ______.

Explanation

Title: ______

I want to explain why (how) ______.

An important reason for why (how) this happens is that ______.

Another reason is that ______.

I know this because ______.

Recursive cycle. An effective method for extending students' science learning through writing is the recursive cycle of research (Bereiter, 2002). This strategy emphasizes writing as a process for learning, similar to the way students learn during the active science investigations.

1. Decide on a problem or question to write about.
2. Formulate an idea or a conjecture about the problem or question.
3. Identify a remedy or an answer, and develop a coherent discussion.
4. Gather information (from an experiment, science notebooks, *FOSS Science Resources*, FOSSweb multimedia, books, Internet, interviews, videos, etc.).
5. Reevaluate the problem or question based on what has been learned.
6. Revise the idea or conjecture.
7. Make presentations (reports, posters, electronic presentations, etc.).
8. Identify new needs, and make new plans.

This process can continue for as long as new ideas and questions occur, or students can present a final product in any of the suggested formats.

CCSS NOTE
This example supports CCSS.ELA-Literacy.W.7.

Narrative Writing

Narrative writing conveys an experience to the reader, usually with sensory detail and a sequence of events. In middle school science, students learn the importance of writing narrative descriptions of their procedures with enough detail and precision to allow others to replicate the experiment. Science also provides a broad landscape of stimulating material for stories, songs, biographies, autobiographies, poems, and plays. Students can enrich their science learning by using organisms or objects as characters; describing habitats and environments as settings; and writing scripts portraying various systems, such as weather patterns, states of matter, and the water, rock, or life cycle.

NOTE
Human characteristics should not be given to organisms (anthropomorphism) in science investigations, only in literacy extensions.

CCSS NOTE
This example supports CCSS.ELA-Literacy.W.3.

READING DOMAIN

Reading is an integral part of science learning. Just as scientists spend a significant amount of their time reading each other's published works, students need to learn to read scientific text—to read effectively for understanding, with a critical focus on the ideas being presented.

CCSS NOTE
The use of *FOSS Science Resources* supports CCSS.ELA-Literacy.RST.6–8.10.

The articles in *FOSS Science Resources* facilitate sense making as students make connections to the science concepts introduced and explored during the active investigations. Concept development is most effective when students are allowed to experience organisms, objects, and phenomena firsthand before engaging the concepts in text. The text and illustrations help students make connections between what they have experienced concretely and the abstract ideas that explain their observations.

FOSS Science Resources provides students with clear and coherent explanations, ways of visualizing important information, and different perspectives to examine and question. As students apply these strategies, they are, in effect, using some of the same scientific thinking processes that promote critical thinking and problem solving. In addition, the text provides a level of complexity appropriate for middle schoolers to develop high-level reading comprehension skills. This development requires support and guidance as students grapple with more complex dimensions of language meaning, structure, and conventions. To become proficient readers of scientific and other academic texts, students must be armed with an array of reading comprehension strategies and have ample opportunities to practice and extend their learning by reading texts that offer new language, new knowledge, and new modes of thought.

CCSS NOTE
Reading breakpoints in the *Investigations Guide* support CCSS.ELA-Literacy.RST.6–8.8.

Oral discourse and writing are critical for reading comprehension and for helping students make sense of the active investigations. Use the suggested prompts, questions, and strategies in the *Investigations Guide* to support comprehension as students read from *FOSS Science Resources*. For most of the investigation parts, the articles are designed to follow the active investigation and are interspersed throughout the course. This allows students to acquire the necessary background knowledge in context through active experience before tackling the wider-ranging content and relationships presented in the text. Breakpoints in the readings are suggested in the *Investigations Guide* to support student conceptual development. Some questions make connections between the reading and the student's class experience. Other questions help the students consider the writer's intent. Additional strategies for reading are derived from the seven essential strategies that readers use to help them understand what they read (Keene and Zimmermann, 2007).

- Monitor for meaning: discover when you know and when you don't know.
- Use and create schemata: make connections between the novel and the known; activate and apply background knowledge.
- Ask questions: generate questions before, during, and after reading that reach for deeper engagement with the text.
- Determine importance: decide what matters most, what is worth remembering.
- Infer: combine background knowledge with information from the text to predict, conclude, make judgments, and interpret.
- Use sensory and emotional images: create mental images to deepen and stretch meaning.
- Synthesize: create an evolution of meaning by combining understanding with knowledge from other texts/sources.

Reading Comprehension Strategies

Below are some strategies that enhance the reading of expository texts in general and have proven to be particularly helpful in science. Read and analyze the articles beforehand in order to guide students through the text structures and content more effectively.

Build on background knowledge. Activating prior knowledge is critical for helping students make connections between what they already know and new information. Reading comprehension improves when students have the opportunity to think, discuss, and write about what they know about a topic before reading. Review what students learned from the active investigation, provide prompts for making connections, and ask questions to help students recall past experiences and previous exposure to concepts related to the reading.

Create an anticipation guide. Create true-or-false statements related to the key ideas in the reading selection. Ask students to indicate if they agree or disagree with each statement before reading, then have them read the text, looking for the information that supports their true-or-false claims. Anticipation guides connect students to prior knowledge, engage them with the topic, and encourage them to explore their own thinking. To provide a challenge for advanced students, have them come up with the statements for the class.

Draw attention to vocabulary. Check the article for bold faced words students may not know. Review the science words that are already defined in students' notebooks. For new science and nonscience

Reading Comprehension Strategies

- *Build on background knowledge*
- *Create an anticipation guide*
- *Draw attention to vocabulary*
- *Preview the text*
- *Turn and talk*
- *Jigsaw text reading*
- *Note making*
- *Summarize and synthesize*
- *3-2-1*
- *Write reflections*
- *Preview and predict*
- *SQ3R*

▶ CCSS NOTE

The example of reviewing what students learned from the active investigation supports CCSS.ELA-Literacy.RST.6–8.9.

▶ CCSS NOTE

This example supports CCSS.ELA-Literacy.RST.6–8.4.

vocabulary words that appear in the reading, have students predict their meanings before reading. During the reading, have students use strategies such as context clues and word structure to see if their predictions were correct. This strategy activates prior knowledge and engages students by encouraging analytical participation with the text.

NOTE
Discourse markers are words or phrases that relate one idea to another. Examples are *however, on the other hand*, and *second.*

CCSS NOTE
This example supports CCSS.ELA-Literacy.RST.6-8.5 and CCSS.ELA-Literacy.RST.6-8.6.

Preview the text. Give students time to skim through the selection, noting subheads, before reading thoroughly. Point out the particular structure of the text and what discourse markers to look for. For example, most *FOSS Science Resources* articles are written as cause and effect, problem and solution, question and answer, comparison and contrast, description, or sequence. Students will have an easier time making sense of the text if they know what text structure to look for. Model and have students practice analyzing these different types of expository text structures by looking for examples, patterns, and discourse markers. For example, let's look at a passage from *FOSS Science Resources: Planetary Science*.

> An eclipse of the Moon occurs when Earth passes exactly between the Moon and the Sun. [cause and effect] The Moon moves into Earth's shadow during a lunar eclipse. At the time of a full lunar eclipse, Earth's shadow completely covers the disk of the Moon. [description] This is how Earth, the Moon, and the Sun are aligned for a lunar eclipse to be observed. [photograph] Why don't we see a lunar eclipse every month? [question and answer] Because of the tilt of the Moon's orbit around the Earth, Earth's shadow does not fall on the Moon in most months.

CCSS NOTE
This example supports CCSS.ELA-Literacy.RST.6-8.7.

Point out how the text in *FOSS Science Resources* is organized (titles, headings, subheadings, questions, and summaries) and if necessary, review how to use the table of contents, glossary, and index. Explain how to scan for formatting features that provide key information (such as boldface type and italics, captions, and framed text) and graphic features (such as tables, graphs, photographs, maps, diagrams, and charts) that help clarify, elaborate, and explain important information in the reading.

While students preview the article, have them focus on the questions that appear in the text, as well as questions at the end of the article. Encourage students to write down questions they have that they think the article will answer.

Turn and talk. When reading as a whole class, stop at key points and have students share their thinking about the selection with the student sitting next to them or with their collaborative group. This strategy helps students process the information and allows everyone to participate in the discussion. When reading in pairs, encourage

students to stop and discuss with their partners. One way to encourage engagement and understanding during paired reading is to have students take turns reading aloud a paragraph or section on a certain topic. The one who is listening then summarizes the meaning conveyed in the passage.

Jigsaw text reading. Students work together in small groups (expert teams) to develop a collective understanding of a text. Each expert team is responsible for one portion of the assigned text. The teams read and discuss their portions to gain a solid understanding of the key concepts. They might use graphic organizers to refine and organize the information. Each expert team then presents its piece to the rest of the class. Or form new jigsaw groups that consist of at least one representative from each expert team. Each student shares with the jigsaw group what their team learned from their particular portion of the text. Together, the participants in the jigsaw group fit their individual pieces together to create a complete picture of the content in the article.

Note making. The more students interact with a reading, the better their understanding. Encourage students to become active readers by asking them to make notes as they read. Studies have shown that note making—especially paraphrasing and summarizing—is one of the most effective means for understanding text (Graham and Herbert, 2010; Applebee, 1984). Some investigation parts include notebook sheets that match pages in *FOSS Science Resources*. This allows students to highlight and underline important points, add notes in the margins, and circle words they do not know. Students can also annotate the article by writing thoughts and questions on self-stick notes. Using symbols or codes can help facilitate comprehension monitoring. Here are some possible symbols students can use to communicate their thinking as they interact with text. (Harvey, 1998).

★ interesting

BK background knowledge

? question

C confusing

I important

L learning something new

W wondering

S surprising

CCSS NOTE
This example supports CCSS.ELA-Literacy.RST.6–8.10.

Students can also use a different set of symbols while making notes about connections: the readings in *FOSS Science Resources* incorporate the active learning that students gain from the investigations, so that they can make authentic text-to-self (T-S) connections. In other words, what they read reminds them of firsthand experiences, making the article more engaging and easier to understand. Text-to-text (T-T) connections are notes students make when they discover a new idea that reminds them of something they've read previously in another text. Text-to-world (T-W) connections involve the text and more global everyday connections to students' lives.

▶ **CCSS NOTE**

This example supports CCSS.ELA-Literacy.RST.6–8.1 and CCSS.ELA-Literacy.RST.6–8.2.

You can model note-making strategies by displaying a selection of text, using a projection system, a document camera, or an interactive whiteboard. As you read the text aloud, model how to write comments on self-stick notes, and use a graphic organizer in a notebook to enhance understanding.

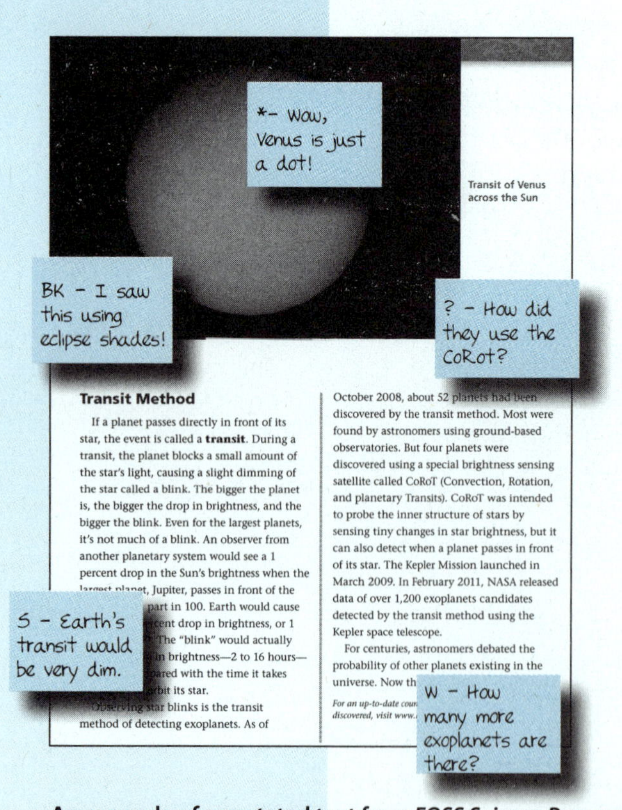

Transit of Venus across the Sun

Transit Method

If a planet passes directly in front of its star, the event is called a **transit**. During a transit, the planet blocks a small amount of the star's light, causing a slight dimming of the star called a blink. The bigger the planet is, the bigger the drop in brightness, and the bigger the blink. Even for the largest planets, it's not much of a blink. An observer from another planetary system would see a 1 percent drop in the Sun's brightness when the largest planet, Jupiter, passes in front of the ... part in 100. Earth would cause ...cent drop in brightness, or 1 ... The "blink" would actually ... in brightness—2 to 16 hours— ...pared with the time it takes ...rbit its star.

Observing star blinks is the transit method of detecting exoplanets. As of October 2008, about 52 planets had been discovered by the transit method. Most were found by astronomers using ground-based observatories. But four planets were discovered using a special brightness sensing satellite called CoRoT (Convection, Rotation, and planetary Transits). CoRoT was intended to probe the inner structure of stars by sensing tiny changes in star brightness, but it can also detect when a planet passes in front of its star. The Kepler Mission launched in March 2009. In February 2011, NASA released data of over 1,200 exoplanets candidates detected by the transit method using the Kepler space telescope.

For centuries, astronomers debated the probability of other planets existing in the universe. Now th...

For an up-to-date coun... discovered, visit www....

FACTS	QUESTIONS
L - When a planet crosses in front of a star, it causes a dimming of the star called a "blink."	Can we observe any other transits in our life time?
I - Exoplanets can be detected by observing star blinks.	How many more exoplanets are there?
C - Earth's transit would cause a very small dimming, but it would be long in brightness.	What does a "long drop in brightness" mean?

An example of annotated text from *FOSS Science Resources: Planetary Science*

Graphic organizers help students focus on extracting the important information from the reading and analyzing relationships between concepts. This can be done by simply having students make columns in their notebooks to record information and their thinking (Harvey and Goudvis, 2007). Here are two examples.

Notes	Thinking

Facts	Questions	Responses

CCSS NOTE
This example supports CCSS.ELA-Literacy.RST.6–8.1 and CCSS.ELA-Literacy.RST.6–8.2.

Summarize and synthesize. Model how to pick out the important parts of the reading selection. Paraphrasing is one way to summarize. Have students write summaries of the reading, using their own words. To scaffold the learning, use graphic organizers to compare and contrast, group, sequence, and show cause and effect. Another method is to have students make two columns in their notebooks. In one column, they record what is important, and in the other, they record their personal responses (what the reading makes them think about). When writing summaries, tell students,

- *Pick out the important ideas.*
- *Restate the main ideas in your own words.*
- *Keep it brief.*

3-2-1. This graphic-organizer strategy gives students the opportunity to synthesize information and formulate questions they still have regarding the concepts covered in an article. In their notebooks, students write three new things they learned, two interesting things worth remembering and sharing, and one question that occurred to them while reading the article. Other options might include three facts, two interesting ideas, and one insight about themselves as learners; three key words, two new ideas, and one thing to think about (modified from Black Hills Special Services Cooperative, 2006).

Write reflections. After reading, ask students to review their notes in their notebooks to make any additions, revisions, or corrections to what they recorded during the reading. This review can be facilitated by using a line of learning. Students draw a line under their original conclusion or under their answer to a question posed at the end of an article. They add any new information as a new narrative entry. The line of learning indicates that what follows represents a change of thinking.

Preview and predict. Instruct students to independently preview the article, directing attention to the illustrations, photos, boldfaced words, captions, and anything else that draws their attention. Working with a partner, students discuss and write three things they think they will learn from the article. Have partners verbally share their list with another pair of students. The group of four can collaborate to generate one list. Groups report their ideas, and together you create a class list on chart paper.

Read the article aloud, or have students read with a partner aloud or silently. Referring to the preview/prediction list, discuss what students learned. Have them record the most important thing they learned from the reading for comparison with the predictions.

SQ3R. Survey, Question, Read, Recall, Reflect strategy provides an overall structure for before, during, and after reading. Students begin by surveying or previewing the text, looking for features that will help them make predictions about the content. Based on their surveys, students develop questions to answer as they read. They read the selections looking for answers to their questions. Next, they recall what they have learned by retelling a partner and/or recording what they've learned. Finally, they reflect on what they have learned, check to see that they've answered their questions sufficiently, and add any new ideas. Below is a chart students can use to record the SQ3R process in their notebooks.

S Survey	Q Question	R Read	R Recall	R Reflect
Scan the text and record important information.	Ask questions about the subject and what you already know.	Record answers to your questions after you read.	Retell what you learned in your own words.	Did you answer your questions? Record new ideas and comments.

Struggling Readers

For students reading below grade level, the strategies listed on the previous pages can be modified to support reading comprehension by integrating scaffolding strategies such as read-alouds and guided reading. Breaking the reading down into smaller chunks, providing graphic organizers, and modeling reading comprehension strategies can also help students who may be struggling with the text. For additional strategies for English learners, see the supported-reading strategy in the English-Language Development section of this chapter.

Struggling Readers
- *Interactive reading aloud*
- *Guided reading*

Interactive reading aloud. Reading aloud is an effective strategy for enhancing text comprehension. It offers opportunities to model specific reading comprehension strategies and allows students to concentrate on making sense of the content. When modeling, share the thinking processes used to understand the reading (questioning, visualizing, comparing, inferring, summarizing, etc.), then have students share what they observed you thinking about as an active reader.

Guided reading. While the rest of the class is reading independently or in small groups, pull a group aside for a guided reading session. Before reading, review vocabulary words from the investigation and ask questions to activate prior knowledge. Have students preview the text to make predictions, ask questions, and think about text structure. Review reading comprehension strategies they will need to use (monitoring for understanding, asking questions, summarizing, synthesizing, etc.). As students read independently, provide support where needed. Ask questions and provide prompts to guide comprehension. (See the list below for additional strategies.) After reading, have students reflect on what strategies they used to help them understand the text and make connections to the investigation.

- While reading, look for answers to questions and confirm predictions.
- Study graphics, such as pictures, graphs, and tables.
- Reread captions associated with pictures, graphs, and tables.
- Note all italicized and boldfaced words or phrases.
- Reduce reading speed for difficult passages.
- Stop and reread parts that are not clear.
- Read only a section at a time, and summarize after each section.

SCIENCE-VOCABULARY DEVELOPMENT

Words play two critically important functions in science. First and most important, we play with ideas in our minds, using words. We present ourselves with propositions—possibilities, questions, potential relationships, implications for action, and so on. The process of sorting out these thoughts involves a lot of internal conversation, internal argument, weighing options, and complex linguistic decisions. Once our minds are made up, communicating that decision, conclusion, or explanation in writing or through verbal discourse requires the same command of the vocabulary. Words represent intelligence; acquiring the precise vocabulary and the associated meanings is key to successful scientific thinking and communication.

The words introduced in FOSS investigations represent or relate to fundamental science concepts and should be taught in the context of the investigation. Many of the terms are abstract and are critical to developing science content knowledge and scientific and engineering practices. The goal is for students to use science vocabulary in ways that demonstrate understanding of the concepts the words represent—not to merely recite scripted definitions. The most effective strategies for science-vocabulary development help students make connections to what they already know. These strategies focus on giving new words conceptual meaning through experience; distinguishing between informal, everyday language and academic language; and using the words in meaningful contexts.

Building Conceptual Meaning through Experience

In most instances, students should be presented with new words when they need to know them in the context of the active experience. Words such as *kinetic energy, atmospheric pressure, chemical reaction, photosynthesis*, and *transpiration* are abstract and conceptually loaded. Students will have a much better chance of understanding, assimilating, and remembering the new word (or new meaning) if they can connect it with a concrete experience.

The vocabulary icon appears in the sidebar when students are prompted to record new words in their notebook. The words that appear in bold are critical to understanding the concepts or scientific practices students are learning and applying in the investigation.

When you introduce a new word, students should

- Hear it: students listen as you model the correct contextual use and pronunciation of the word;
- See it: students see the new word written out;
- Say it: students use the new word when discussing their observations and inferences; and
- Write it: students use the new words in context when they write in their notebooks.

Bridging Informal Language to Science Vocabulary

FOSS investigations are designed to tap into students' inquisitive nature and their excitement of discovery in order to encourage lively discussions as they explore materials in creative ways. There should be a lot of talking during the investigations! Your role is to help students connect informal language to the vocabulary used to express specific science concepts. As you circulate during active investigations, you continually model the use of science vocabulary. For example, as students are examining a leaf under the microscope, they will say, "I can see little mouths." You might respond, "Yes, those mouthlike openings are called stomates. They are pores that open and close." Below are some strategies for validating students' conversational language while developing their familiarity with and appreciation for science vocabulary.

Bridging Informal Language to Science Vocabulary

- *Cognitive-content dictionaries*
- *Concept maps*
- *Semantic webs*
- *Word associations*
- *Word sorts*

Cognitive-content dictionaries. Choose a term that is critical for conceptual understanding of the science investigation. Have students write the term, predict its meaning, write the final meaning after class discussion (using primary language or an illustration), and use the term in a sentence.

Cognitive-Content Dictionary	
New term	kinetic energy
Prediction (clues)	something that moves a lot
Final meaning	motion energy
How I would use it in a sentence	Fast-moving particles have more kinetic energy than slow-moving particles.

Concept maps. Select six to ten related science words. Have students write them on self-stick notes or cards. Have small groups discuss how the words are related. Students organize words in groups and glue them down or copy them on a sheet of paper. Students draw lines between the related words. On the lines, they write words describing or explaining how the concept words are related.

Semantic webs. Select a vocabulary word, and write it in the center of a piece of paper (or on the board for the whole class). Brainstorm a list of words or ideas that are related to the first word. Group the words and concepts into several categories, and attach them to the central word with lines, forming a web (modified from Hamilton, 2002).

Word associations. In this brainstorming activity, you say a word, and students respond by writing the first word that comes to mind. Then students share their words with the class. This activity builds connections to students' prior frames of reference.

Word sorts. Have students work with a partner to make a set of word cards using new words from the investigation. Have them group the words in different ways, for example, synonyms, root words, and conceptual connections.

Science Vocabulary Strategies

- *Word wall*
- *Drawings and diagrams*
- *Cloze activity*
- *Word wizard*
- *Word analysis/word parts*
- *Breaking apart words*
- *Possible sentences*
- *Reading*
- *Glossary*
- *Index*
- *Poems, chants, and songs*

Using Science Vocabulary in Context

For a new vocabulary word to become part of a student's functional vocabulary, he or she must have ample opportunities to hear and use it. Vocabulary terms are used in the activities through teacher talk, whole-class and small-group discussions, writing in science notebooks, readings, and assessments. Other methods can also be used to reinforce important vocabulary words and phrases.

Word wall. Use chart paper to record science content and procedural words as they come up during and after the investigations. Students will use this word wall as a reference.

Drawings and diagrams. For English learners and visual learners, use a diagram to review and explain abstract content. Ahead of time, draw an illustration lightly, almost invisibly, with pencil on chart paper. You can do this easily by projecting the image onto the paper. When it's time for the investigation, trace the illustration with markers as you introduce the words and phrases to students. Students will be amazed by your artistic ability.

Cloze activity. Structure sentences for students to complete, leaving out the vocabulary words. This can be done as a warm-up with the words from the previous day's lesson. Here's an example from the **Earth History Course**.

Teacher: *The removal and transportation of loose earth materials is called ______.*

Students: *Erosion.*

Word wizard. Tell students that you are going to lead a word activity. You will be thinking of a science vocabulary word. The goal is to figure out the word. Provide hints that have to do with parts of a definition, root word, prefix, suffix, and other relevant components. Students work in teams of two to four. Provide one hint, and give teams 1 minute to discuss. One team member writes the word on a piece of paper or on the whiteboard, using dark marking pens. Each team holds up its word for only you to see. After the third clue, reveal the word, and move on to the next word. Here's an example.

1. *Part of the word means green.*
2. *They are found in plant cells.*
3. *They look like tiny green spheres or ovals.*

The word is ***chloroplasts****.*

Word analysis/word parts. Learning clusters of words that share a common origin can help students understand content-area texts and connect new words to familiar ones. Here's an example: *geology, geologist, geological, geography, geometry, geophysical.* This type of contextualized teaching meets the immediate need of understanding an unknown word while building generative knowledge that supports students in figuring out difficult words for future reading.

Breaking apart words. Have teams of two to four students break a word into prefix, root word, and suffix. Give each team different words, and have each team share the parsed elements of the word with the whole class. Here's an example.

photosynthesis

Prefix = *photo*: meaning light

Root = *synthesis*: meaning to put together

Possible sentences. Here is a simple strategy for teaching word meanings and generating class discussion.

1. Choose six to eight key concept words from the text of an article in *FOSS Science Resources*.
2. Choose four to six additional words that students are more likely to know something about.
3. Put the list of ten to fourteen words on the board or project it. Provide brief definitions as needed.
4. Ask students to devise sentences that include two or more words from the list.
5. On chart paper, write all sentences that students generate, both coherent and otherwise.
6. Have students read the article from which the words were extracted.
7. Revisit students' sentences, and discuss whether the sentences are sensible based on the passage or how they could be modified to be more coherent.

Reading. After the active investigation, students continue to develop their understanding of the vocabulary words and the concepts those words represent by listening to you read aloud, reading with a partner, or reading independently. Use strategies discussed in the Reading Domain section to encourage students to articulate their thoughts and practice the new vocabulary.

Glossary. Emphasize the vocabulary words students should be using when they answer the focus question in their science notebooks. The glossary in *FOSS Science Resources* or on FOSSweb can be used as a reference.

NOTE
See the Science Notebooks in Middle School chapter for an example of an index.

Index. Have students create an index at the back of their notebooks. There they can record new vocabulary words and the notebook page where they defined and used the new words for the first time in the context of the investigation.

Poems, chants, and songs. As extensions or homework assignments, ask students to create poems, raps, chants, or songs, using vocabulary words from the investigation.

ENGLISH-LANGUAGE DEVELOPMENT

Active investigations, together with ample opportunities to develop and use language, provide an optimal learning environment for English learners. This section highlights opportunities for English-language development (ELD) in FOSS investigations and suggests other best practices for facilitating both the learning of new science concepts and the development of academic literacy. For example, the hands-on structure of FOSS investigations is essential for the conceptual development of science content knowledge and the habits of mind that guide and define scientific and engineering practices. Students are engaged in concrete experiences that are meaningful and that provide a shared context for developing understanding—critical components for language acquisition.

When getting ready for an investigation, review the steps and determine the points where English learners may require scaffolds and where the whole class might benefit from additional language-development supports. One way to plan for ELD integration in science is to keep in mind four key areas: prior knowledge, comprehensible input, academic language development, and oral practice. The ELD chart lists examples of universal strategies for each of these components that work particularly well in teaching science.

NOTE
English-language development refers to the advancement of students' ability to read, write, and speak English.

English-Language Development (ELD)	
Activating prior knowledge • Inquiry chart • Circle map • Observation poster • Quick write • Kit inventory	**Using comprehensible input** • Content objectives • Multiple exposures • Visual aids • Supported reading • Procedural vocabulary
Developing academic language • Language objectives • Sentence frames • Word wall, word cards, drawings • Concept maps • Cognitive content dictionaries	**Providing oral practice** • Small-group discussions • Science talk • Oral presentations • Poems, chants, and songs • Teacher feedback

NOTE
Language forms and structures are the internal grammatical structure of words and how those words go together to make sentences.

Students acquiring English benefit from scaffolds that support the language forms and functions necessary for the academic demands of the science course, that is, accessing science text, participating in productive oral discourse, and engaging in science writing. The table at the end of this section (starting on page 38) provides a resource to help students organize their thinking and structure their speaking and writing in the context of the science and engineering practices. The table identifies key language functions exercised during FOSS investigations and provides examples of sentence frames students can use as scaffolds.

For example, if students are planning an investigation to learn more about insect structures and behaviors, the language objective might be "Students plan and design an investigation that answers a question about the hissing cockroach's behavior." For students who need support, a sentence frame that prompts them to identify the variables in the investigation would provide language forms and structures appropriate for planning their investigation. As a scaffold, sentence frames can also help them write detailed narratives of their procedure. Here's an example from the table.

NOTE
The complete table appears at the end of the English-Language Development section starting on page 38.

Language functions	Language objectives	Sentence frames
Planning and carrying out investigations		
Design Sequence Strategize Evaluate	Plan controlled experiments with multiple trials. Identify independent variable and dependent variable. Discuss, describe, and evaluate the methods for collecting data.	To find out ______, I will change ______. I will not change ______. I will measure ______. I will observe ______. I will record the data by ______. First, I will ______, and then I will ______. To learn more about ______, I will need ______ to ______.

Activating Prior Knowledge

When an investigation engages a new concept, students first recall and discuss familiar situations, objects, or experiences that relate to and establish a foundation for building new knowledge and conceptual understanding. Eliciting prior knowledge also supports learning by motivating interest, acknowledging culture and values, and checking for misconceptions and prerequisite knowledge. This is usually done in the first steps of Guiding the Investigation in the form of a discussion, presentation of new materials, or a written response to a prompt. The tools outlined below can also be used before beginning an investigation to establish a familiar context for launching into new material.

Circle maps. Draw two concentric circles on chart paper. In the smaller circle, write the topic to be explored. In the larger circle, record what students already know about the subject. Ask students to think about how they know or learned what they already know about the topic. Record the responses outside the circles. Students can also do this independently in their science notebooks.

An example of a circle map

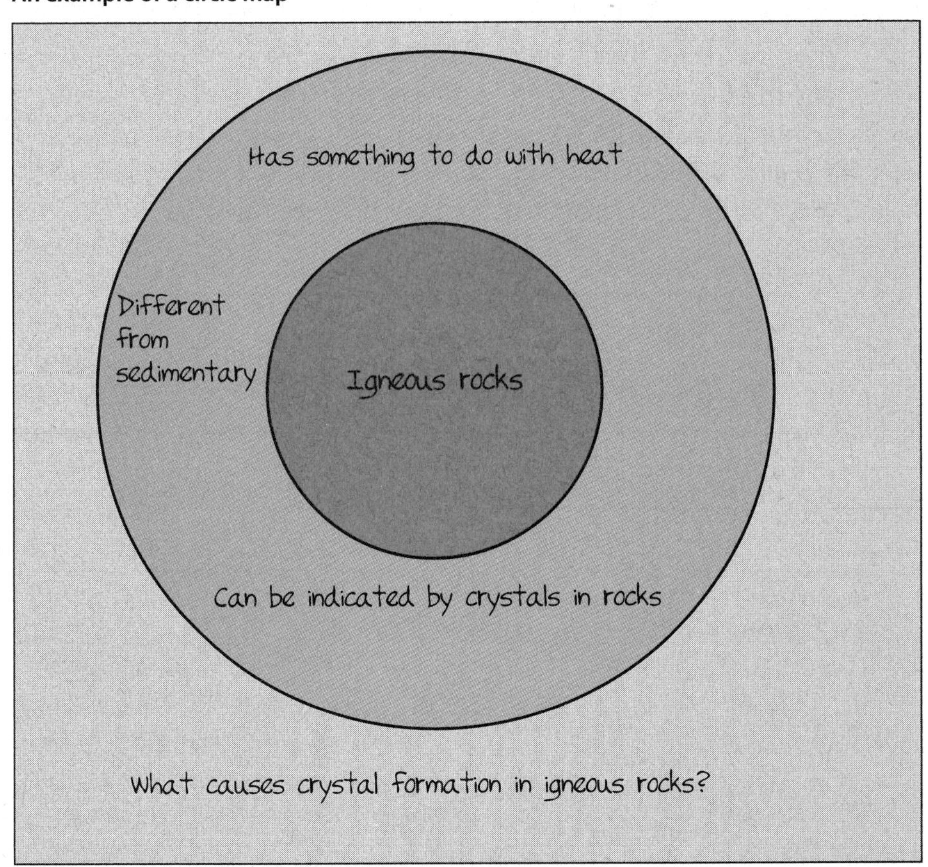

Activating Prior Knowledge

- *Circle maps*
- *Observation posters*
- *Quick writes*
- *Kit inventories*

Observation posters. Make observation posters by gluing or taping pictures and artifacts relevant to the module or a particular investigation onto pieces of blank chart paper or poster paper. Hang them on the wall in the classroom, and have students rotate in small groups to each poster. At each poster, students discuss their observations with their partners or small groups and then record (write or draw) an observation, a question, a prediction, or an inference about the pictures as a contribution to the commentary on the poster.

As a variation on this strategy, give a set of pictures to each group to pass around. Have them choose one and write what they notice, what they infer, and questions they have in their notebooks.

Quick writes. Ask students what they know about the topic of the investigation. Responses can be recorded independently as a quick write in notebooks and then shared collaboratively. Do not correct misconceptions initially. Periodically revisit the quick-write ideas as a whole class, or have students review their notebook entries to correct, confirm, or complete their original thoughts as they acquire new information (possibly using a line of learning). At the conclusion of the investigation, students should be able to express mastery of the new conceptual material.

Kit inventories. Introduce each item from the FOSS kit used in the investigation, and ask students questions to get them thinking about what each item is and where they may have seen it before. Have them describe the objects and predict how they will be used in the investigation.

Comprehending Input

To initiate their own sense making, students must be able to access the information presented to them. We refer to this ability as comprehending input. Students must understand the essence of new ideas and concepts before beginning to construct new scientific meaning. The strategies for comprehensible input used in FOSS ensure that the instruction is understandable while providing students with the opportunity to grapple with new ideas and the critically important relationships between concepts. Additional tools such as repetition, visual aids, emphasis on procedural vocabulary, and auditory reinforcement can also be used to enhance comprehensible input for English learners.

Content objectives. The focus question for each investigation part frames the activity objectives—what students should know or be able to do at the end of the part. Making the learning objectives clear and explicit prepares English learners to process the delivery of new information, and helps you maintain the focus of the investigation. Write the focus question on the board, have students read it aloud and transcribe it into their science notebooks, and have students answer the focus question at the end of the investigation part. You then check their responses for understanding.

Multiple exposures. Repeat an activity in an analogous but slightly different context, ideally one that incorporates elements that are culturally relevant to students. For example, as a homework assignment for landforms, have students interview their parents about landforms common in the area of their ancestry.

Visual aids. On the board or chart paper, write out the steps for conducting the investigation. This provides a visual reference. Include illustrations if necessary. Use graphic representations (illustrations drawn and labeled in front of students) to review the concepts explored in the active investigations. In addition to the concrete objects in the kit, use realia to augment the activity, to help English learners build understanding and make cultural connections. Graphic organizers (webs, Venn diagrams, T-tables, flowcharts, etc.) aid comprehension by helping students see how concepts are related.

Supported reading. In addition to the reading comprehension strategies suggested in the Reading Domain section of this chapter, English learners can also benefit from methods such as front-loading key words, phrases, and complex text structures before reading; using

Comprehending Input

- *Content objectives*
- *Multiple exposures*
- *Visual aids*
- *Supported reading*
- *Procedural vocabulary*

Procedural Vocabulary

Add
Analyze
Assemble
Attach
Calculate
Change
Classify
Collect
Communicate
Compare
Connect
Construct
Contrast
Demonstrate
Describe
Determine
Draw
Evaluate
Examine
Explain
Explore
Fill
Graph
Identify
Illustrate
Immerse
Investigate
Label
List
Measure
Mix
Observe
Open
Order
Organize
Pour
Predict
Prepare
Record
Represent
Scratch
Separate
Sort
Stir
Subtract
Summarize
Test
Weigh

preview-review (main ideas are previewed in the primary language, read in English, and reviewed in the primary language); and having students use sentence frames specifically tailored to record key information and/or graphic organizers that make the content and the relationship between concepts visually explicit from the text as they read.

Procedural vocabulary. Make sure students understand the meaning of the words used in the directions for an investigation. These may or may not be science-specific words. Use techniques such as modeling, demonstrating, and body language (gestures) to explain procedural meaning in the context of the investigation. The words students will encounter in FOSS include those listed in the sidebar. To build academic literacy, English learners need to learn the multiple meanings of these words and their specific meanings in the context of science.

Developing Academic Language

As students learn the nuances of the English language, it is critical that they build proficiency in academic language in order to participate fully in the cognitive demands of school. *Academic language* refers to the more abstract, complex, and specific aspects of language, such as the words, grammatical structure, and discourse markers that are needed for higher cognitive learning. FOSS investigations introduce and provide opportunities for students to practice using the academic vocabulary needed to access and engage with science ideas.

Language objectives. Consider the language needs of English learners and incorporate specific language-development objectives that will support learning the science content of the investigation, such as a specific way to expand use of vocabulary by looking at root words, prefixes, and suffixes; a linguistic pattern or structure for oral discussion and writing; or a reading comprehension strategy. Recording in science notebooks is a productive way to optimize science learning and language objectives. For example, in the **Earth History Course**, one language objective might be "Students will apply techniques for rock observations to compare and contrast sedimentary and igneous rocks. They will discuss and record their observations in their notebooks in an organized manner."

Vocabulary development. The Science-Vocabulary Development section in this chapter describes the ways science vocabulary is introduced and developed in the context of an active investigation and suggests methods and strategies that can be used to support vocabulary development during science instruction. In addition to science vocabulary, students need to learn the nonscience vocabulary that facilitates deeper understanding and communication skills. Words such as *release*, *convert*, *beneficial*, *produce*, *receive*, *source*, and *reflect* are used in the investigations and *FOSS Science Resources* and are frequently used in other content areas. Learning these academic-vocabulary words gives students a more precise and complex way of practicing and communicating productive thinking. Consider using the strategies described in the Science-Vocabulary Development section to explicitly teach targeted, high-leverage words that can be used in multiple ways and that can help students make connections to other words and concepts. Sentence frames, word walls, concept maps, and cognitive-content dictionaries are strategies that have been found to be effective with academic-vocabulary development.

Scaffolds That Support Science and Engineering Practices

Language functions	Language objectives	Sentence frames
Asking questions and defining problems		
Inquire Define a problem	Ask questions to solicit information about phenomena, models, or unexpected results; determine the constraints and criteria of a problem.	I wonder why ____ . What happens when ____? What if ____? What does ____? What can ____? What would happen if ____? How does ____ affect ____? How can I find out if ____? Which ____ is better for ____?
Planning and carrying out investigations		
Design Sequence Strategize Evaluate	Plan controlled experiments with multiple trials. Identify independent variable and dependent variable. Discuss, describe, and evaluate the methods for collecting data.	To find out ____, I will change ____. I will not change ____. I will measure ____. I will observe ____. I will record the data by ____. First, I will ____, and then I will ____. To learn more about ____, I will need ____ to ____.

Language functions	Language objectives	Sentence frames
Planning and carrying out investigations *(continued)*		
Describe	Write narratives using details to record sensory observations and connections to prior knowledge.	I observed/noticed ____. When I touch the ____, I feel ____. It smells ____. It sounds ____. It reminds me of ____, because ____.
Organize Compare Classify	Make charts and tables: use a T-table or chart for recording and displaying data.	The table compares ____ and ____
Sequence Compare	Record changes over time, and describe cause-and-effect relationships.	At first, ____, but now ____. We saw that first ____, then ____, and finally ____. When I ____, it ____. After I ____, it ____.
Draw Label Identify	Draw accurate and detailed representations; identify and label parts of a system using science vocabulary, with attention to form, location, color, size, and scale.	The diagram shows ____. ____ is shown here. ____ is ____ times bigger than ____. ____ is ____ times smaller than ____ .
Analyzing and interpreting data		
Enumerate Compare Represent	Use measures of variability to analyze and characterize data; decide when and how to use bar graphs, line plots, and two-coordinate graphs to organize data.	The mean is ____. The median is ____. The mode is ____. The range is ____. The x-axis represents ____ and the y-axis represents ____. The units are expressed in ____.

Language functions	Language objectives	Sentence frames
Analyzing and interpreting data *(continued)*		
Compare Classify Sequence	Use graphic organizers and narratives to express similarities and differences, to assign an object or action to the category or type to which it belongs, and to show sequencing and order.	This _____ is similar to _____ because _____. This _____ is different from _____ because _____. All these are _____ because _____. _____, _____, and _____ all have/are _____.
Analyze	Use graphic organizers, narratives, or concept maps to identify part/whole or cause-and-effect relationships. Express data in qualitative terms such as more/fewer, higher/lower, nearer/farther, longer/shorter, and increase/decrease; and quantitatively in actual numbers or percentages.	The _____ consists of _____. The _____ contains _____. As _____, then _____. When I changed _____, then _____ happened. The more/less _____, then _____.
Developing and using models		
Represent Predict Explain	Construct and revise models to predict, represent, and explain.	If _____, then _____, therefore _____. The _____ represents _____ . _____ shows how _____. You can explain _____ by _____.

Language functions	Language objectives	Sentence frames
Using mathematics and computational thinking		
Symbolize Measure Enumerate Estimate	Use mathematical concepts to analyze data.	The ratio of ___ is ___ to ___. The average is ___. Looking at ___, I think there are ___. My prediction is ___.
Constructing explanations and designing solutions		
Infer Explain	Construct explanations based on evidence from investigations, knowledge, and models; use reasoning to show why the data are adequate for the explanation or conclusion.	I claim that ___. I know this because ___. Based on ___, I think ___. As a result of ___, I think ___. The data show ___, therefore, ___. I think ___ means ___ because ___. I think ___ happened because ___.
Provide evidence	Use qualitative and quantitative data from the investigation as evidence to support claims. Use quantitative expressions using standard metric units of measurement such as cm, mL, °C.	My data show ___. My evidence is ___. The relationship between the variables is ___. The model of ___ shows that ___.

<table>
<tr><th>Language functions</th><th>Language objectives</th><th>Sentence frames</th></tr>
<tr><th colspan="3">Engaging in argument from evidence</th></tr>
<tr><td>Discuss
Persuade
Synthesize
Negotiate
Suggest</td><td>Use oral and written arguments supported by evidence and reasoning to support or refute an argument for a phenomenon or a solution to a problem.</td><td>I think ___ because___.
I agree/disagree with ___ because__________.
What you are saying is ______. What do you think about _____?
What if ______?
I think you should try ___.
Another way to interpret the data is ______.</td></tr>
<tr><td>Critique
Evaluate
Reflect</td><td>Evaluate competing design solutions based on criteria; compare two arguments from evidence to identify which is better.</td><td>____ makes more sense because ____.
____ is a better design _____ because it ____.
Comparing ___ to ___ shows that ______.
One discrepancy is ____.
____ is inconsistent with _____.
Another way to determine ______ is to _______.
I used to think ____, but now I think ____. I have changed my thinking about ____.
I am confused about ____ because ____.
I wonder ____.</td></tr>
<tr><th colspan="3">Obtaining, evaluating, and communicating information</th></tr>
<tr><td colspan="3">(This practice includes all functions described in the other practices above.)</td></tr>
</table>

REFERENCES

Applebee, A. 1984. "Writing and Reasoning." *Review of Educational Research* 54 (Winter): 577–596.

Bereiter, C. 2002. *Education and Mind in the Knowledge Age.* Hillsdale, NJ: Erlbaum.

Black Hills Special Services Cooperative. 2006. "3-2-1 Strategy." In *On Target: More Strategies to Guide Learning.* Rapid City, SD: South Dakota Department of Education.

Gibbons, P. 2002. *Scaffolding Language, Scaffolding Learning.* Portsmouth, NH: Heinemann.

Graham, S., and M. Herbert. 2010. *Writing to Read: Evidence for How Writing Can Improve Reading.* New York: Carnegie.

Hamilton, G. 2002. *Content-Area Reading Strategies: Science.* Portland, ME: Walch Publishing.

Harvey, S. 1998. *Nonfiction Matters: Reading, Writing, and Research in Grades 3–8.* Portland, ME: Stenhouse.

Harvey, S., and A. Goudvis. 2007. *Strategies That Work: Teaching Comprehension for Understanding and Engagement.* Portland, ME: Stenhouse.

Keeley, P. 2008. *Science Formative Assessment: 75 Practical Strategies for Linking Assessment, Instruction, and Learning.* Thousand Oaks, CA: Corwin Press.

Keene, E., and S. Zimmermann. 2007. *Mosaic of Thought: The Power of Comprehension Strategies.* 2nd ed. Portsmouth, NH: Heinemann.

Keys, C. 1999. *Revitalizing Instruction in Scientific Genres: Connecting Knowledge Production with Writing to Learn in Science.* Athens: University of Georgia.

Lieberman, G. A., and L. L. Hoody. 1998. *Closing the Achievement Gap: Using the Environment as an Integrating Context for Learning.* San Diego, CA: State Education and Environment Roundtable.

National Governors Association Center for Best Practices, Council of Chief State School Officers. 2010. *Common Core State Standards for English Language Arts & Literacy in History/Social Studies, Science, and Technical Subjects.* Washington, DC: National Governors Association Center for Best Practices, Council of Chief State School Officers.

National Research Council. 2012. *A Framework for K–12 Science Education: Practices, Crosscutting Concepts, and Core Ideas.* Committee on a Conceptual Framework for New K–12 Science Education Standards. Board on Science Education, Division of Behavioral and Social Sciences and Education. Washington, DC: The National Academies Press.

NOTE

For additional resources and updated references, go to FOSSweb.

Norris, S. P., and L. M. Phillips. 2003. "How Literacy in Its Fundamental Sense Is Central to Scientific Literacy." *Science Education* 87 (2).

Ostlund, K. 1998. "What the Research Says about Science Process Skills: How Can Teaching Science Process Skills Improve Student Performance in Reading, Language Arts, and Mathematics?" *Electronic Journal of Science Education* 2 (4).

Wellington, J., and J. Osborne. 2001. *Language and Literacy in Science Education*. Buckingham, UK: Open University Press.

Winokur, J., and K. Worth. 2006. "Talk in the Science Classroom: Looking at What Students and Teachers Need to Know and Be Able to Do." In *Linking Science and Literacy in the K–8 Classroom*, ed. R. Douglas, K. Worth, and W. Binder. Arlington, VA: NSTA Press.

FOSSweb and Technology

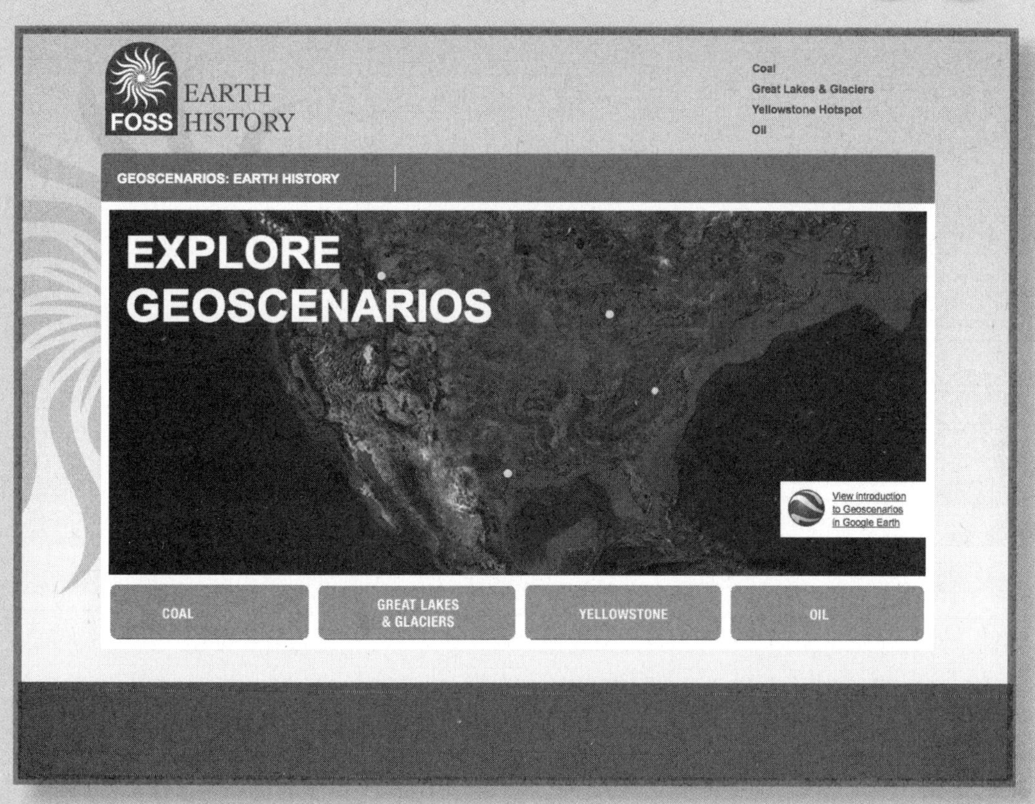

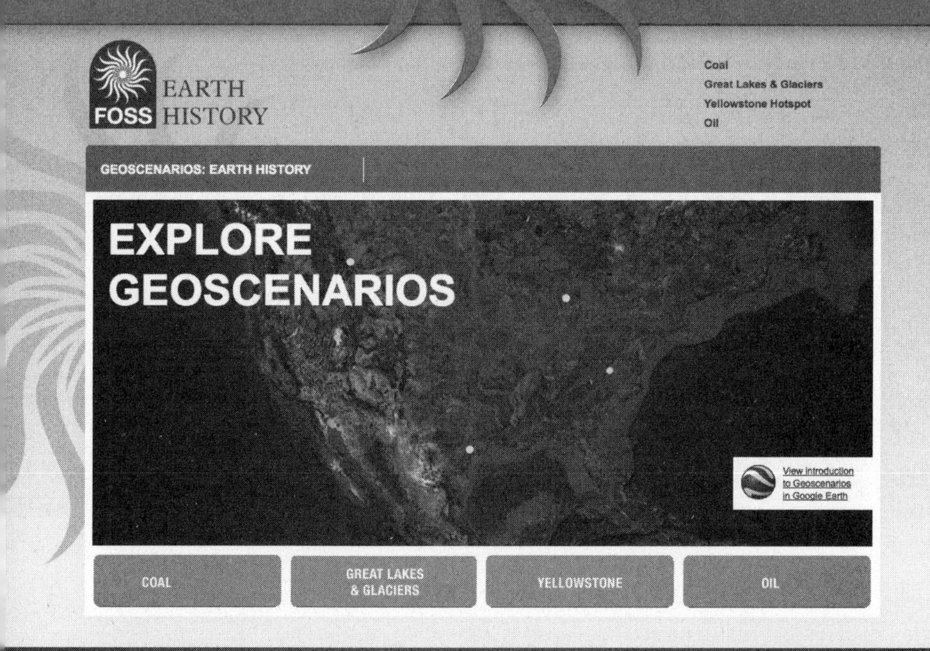

Contents

INTRODUCTION

FOSSweb technology is an integral part of the **Earth History Course**. It provides students with the opportunity to access and interact with simulations, images, video, and text—digital resources that can enhance their understanding of life science concepts. Different sections of digital resources are incorporated into each investigation during the course. Each use is marked with the technology icon in the *Investigations Guide*. You will sometimes use the digital resources to make presentations to the class. At other times, individuals or small groups of students will work with the digital resources to review concepts or reinforce their understanding.

The FOSSweb components are not optional. To prepare to use these digital resources, you should have at a minimum one device with Internet access that can be displayed to the class by an LCD projector with an interactive whiteboard or a large screen arranged for class viewing. Access to a computer lab or to enough devices in your classroom for students to work in small groups is also required during one investigation, and recommended during others.

The digital resources are available online at www.FOSSweb.com for teachers and students. We recommend you access FOSSweb well in advance of starting the course to set up your teacher-user account and become familiar with the resources.

REQUIREMENTS FOR ACCESSING FOSSWEB

You'll need to have a few things in place on your device before accessing FOSSweb. Once you're online, you'll create a FOSSweb account. All information in this section is updated as needed on FOSSweb.

Creating a FOSSweb Teacher Account

By creating a FOSSweb teacher account, you can personalize FOSSweb for easy access to the courses you are teaching. When you log in, you will be able to add courses to your "My FOSS Modules" area and access Resources by Investigation for the **Earth History Course**. This makes it simple to select the investigation and part you are teaching and view all the digital resources connected to that part.

Students and families can also access course resources through FOSSweb. You can set up a class account and class pages where students will be able to access notes from you about assignments and digital resources.

Setting up an account. Set up an account on FOSSweb so you can access the site when you begin teaching a course. Go to FOSSweb to register for an account—complete registration instructions are available online.

Entering your access code. Once your account is set up, go to FOSSweb and log in. The first time you log in, you will need to enter your access code. Your access code should be printed on the inside cover of your *Investigations Guide*. If you cannot find your FOSSweb access code, contact your school administrator, your district science coordinator, or the purchasing agent for your school or district.

Familiarize yourself with the layout of the site and the additional resources available by using your account login. From your course page, you will be able to access teacher masters, assessment masters, notebook sheets, and other digital resources.

Explore the Resources by Investigation section, as this will help you plan. It provides links to notebook sheets, teacher masters, online activities, and the teaching slides for each investigation part.

There are also a variety of beneficial resources on FOSSweb that can be used to assist with teacher preparation and materials management including teacher preparation videos for each investigation, editable teaching slides to use in conducting the investigations, and a planning guide to help you set up your classroom for three-dimensional teaching and learning.

Setting up class pages and student accounts. To enable your students to log in to FOSSweb to access the digital resources and see class assignments, set up a class page and generate a user name and password for the class. To do so, log in to FOSSweb and go to your teacher homepage. Under My Class Pages, follow the instructions to create a new class page and to leave notes for students.

If a class page and student accounts are not set up, students can always access digital resources by visiting FOSSweb.com and choosing to visit the site as a guest.

FOSSweb Technical Requirements

To use FOSSweb, your device must meet minimum system requirements and have a compatible browser and recent versions of any required plug-ins. The system requirements are subject to change. You must visit FOSSweb to review the most recent minimum system requirements.

Preparing your browser. FOSSweb requires a supported operating system with current versions of all required plug-ins. You may need administrator privileges on your device in order to install the required programs and/or help from your school's technology coordinator. Check compatibility for each device you will use to access FOSSweb by accessing the Technical Requirements page on FOSSweb. The information at FOSSweb contains the most up-to-date technical requirements.

https://www.FOSSweb.com/tech-specs-and-info

Support for plug-ins and reader. Any required Adobe plug-ins are available on www.Adobe.com as free downloads. If required, QuickTime is available free of charge from www.Apple.com. FOSS does not support these programs. Please go to the program's website for troubleshooting information.

Accessing FOSS Earth History Digital Resources

When you log in to FOSSweb, the most useful way to access course materials on a daily basis is the Resources by Investigation section of the **Earth History Course** page. This section lists the digital resources, student and teacher sheets, readings, and focus questions for each investigation part. Each of these items is linked so you can click and go directly to that item.

Students will access digital resources from the Resource Room, accessible from the class page you've set up. Explore where the activities reside in the Resource Room. At various points in the course, students will access interactive simulations, images, videos, and animations from FOSSweb.

Other Technology Considerations

Firewall or proxy settings. If your school has a firewall or proxy server, contact your IT administrator to add explicit exceptions in your proxy server and firewall for these servers:

- fossweb.com
- fossweb.schoolspecialty.com
- archive.fossweb.com
- science.video.schoolspecialty.com
- a445.w10.akamai.net

Classroom technology setup. FOSS has a number of digital resources and makes every effort to accommodate users with different levels of access to technology. The digital resources can be used in a variety of ways and can be adapted to a number of classroom setups.

Teachers with classroom devices and an LCD projector, an interactive whiteboard, or a large screen will be able to show multimedia to the class. If you have access to a computer lab, or enough devices in your classroom for students to work in small groups, you can set up time for students to use the FOSSweb digital resources during the school day.

Displaying digital content. You might want to digitally display the notebook and teacher masters during class. In the Resources by Investigation section of FOSSweb, you'll have the option of downloading the masters "to project" or "to copy." Choose "to project" if you plan on projecting the masters to the class. These masters are optimized for a projection system and allow you to type into them while they are displayed. The "to copy" versions are sized to minimize paper use when photocopying for the class, and to fit optimally into student notebooks.

NOTE
FOSSweb activities are designed for a minimum screen size of 1024 × 768. It is recommended that you adjust your screen resolution to 1024 × 768 or higher.

TROUBLESHOOTING AND TECHNICAL SUPPORT

If you experience trouble with FOSSweb, you can troubleshoot in a variety of ways.

1. Visit FOSSweb and make sure your devices meet the minimum system requirements

   https://www.FOSSweb.com/tech-specs-and-info

2. Check the FAQs on FOSSweb for additional information that may help resolve the problem.
3. Try emptying the cache from your browser and/or quitting and relaunching it.
4. Restart your device, and make sure all hardware turns on and is connected correctly.
5. If your school has a firewall or proxy server, contact your IT administrator to add explicit exceptions in your proxy server and firewall for the servers listed on the previous page in this chapter.

If you are still experiencing problems after taking these steps, send FOSS Technical Support an e-mail at support@FOSSweb.com. In addition to describing the problem, include the following information about your device: name of device, operating system, browser and version, plug-ins and versions. This will help us troubleshoot the problem.

Science Notebook Masters

Landforms Tour

	Site name	Landforms, observations, and questions
1	My school	
2	Mount St. Helens	
3	Aleutian Islands	
4	Alaskan glaciers	
5	Na Pali Coast	
6	Death Valley	
7	Florida Keys	
8	Stone Mountain	
9	Shenandoah	
10	Central Park	
11	Niagara Falls	
12	Delta	
13	Mississippi River	
14	Lake of the Ozarks	
15	Oklahoma Panhandle	
16	Missouri River	
17	Grand Tetons	
18	Grand Canyon	

Landforms Tour

	Site name	Landforms, observations, and questions
1	My school	
2	Mount St. Helens	
3	Aleutian Islands	
4	Alaskan glaciers	
5	Na Pali Coast	
6	Death Valley	
7	Florida Keys	
8	Stone Mountain	
9	Shenandoah	
10	Central Park	
11	Niagara Falls	
12	Delta	
13	Mississippi River	
14	Lake of the Ozarks	
15	Oklahoma Panhandle	
16	Missouri River	
17	Grand Tetons	
18	Grand Canyon	

FOSS Next Generation
© The Regents of the University of California
Can be duplicated for classroom or workshop use.

Earth History Course
Investigation 1: Earth Is Rock
No. 1—Notebook Master

Landform Vocabulary

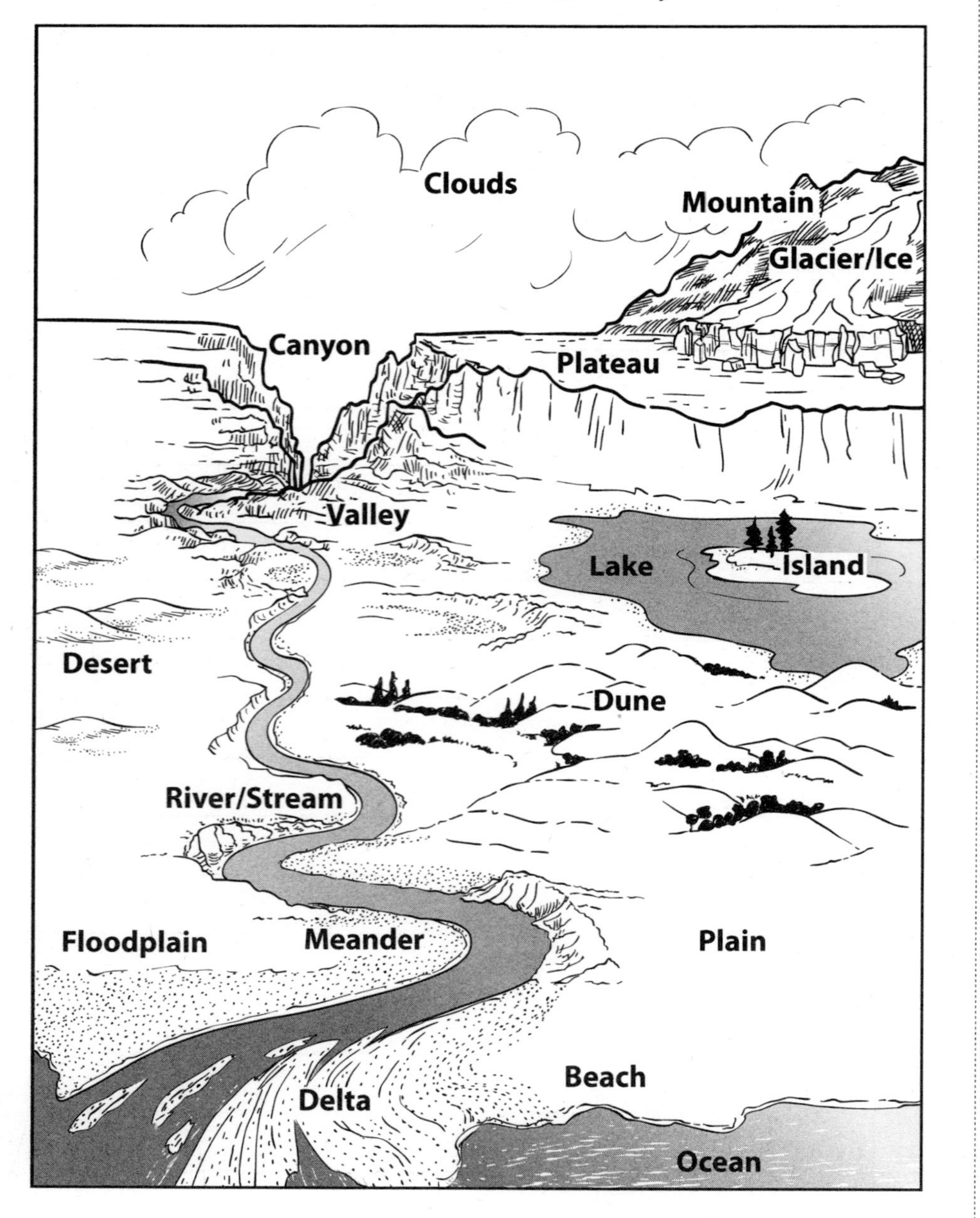

FOSS Next Generation

Landform Vocabulary

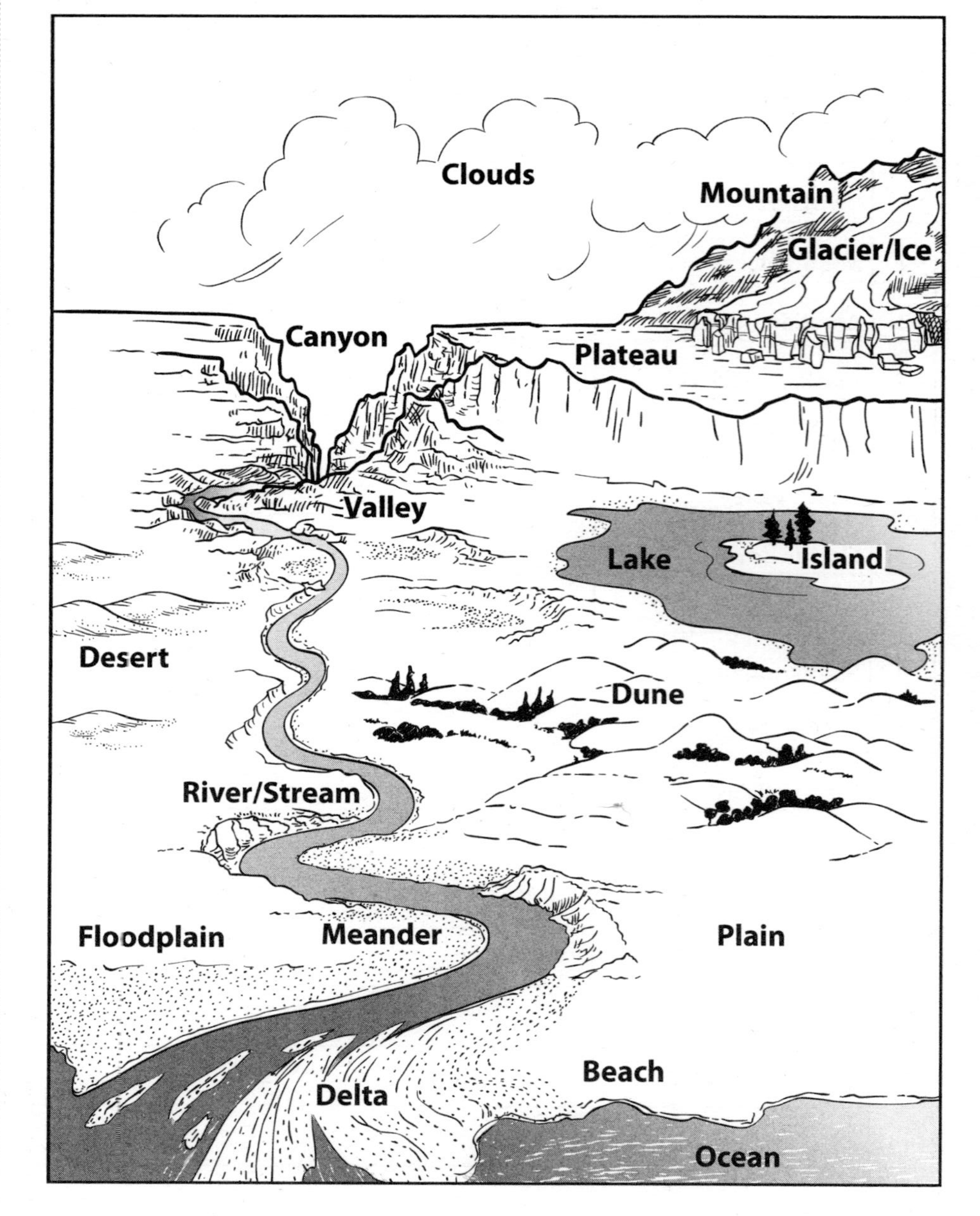

FOSS Next Generation

Anticipation Guide

Before reading the article "Seeing Earth," check off the statements with which you agree.

____ 1. Images in Google Earth™ show live images of Earth as it appears right now.

____ 2. Rocks are actual pieces of Earth.

____ 3. Google Earth™ shows representations of the surface of planet Earth.

____ 4. Earth-imaging satellites can take a picture of the whole surface of one side of Earth.

After reading the article, check off the statements that were confirmed in the text.

____ 1. Images in Google Earth™ show live images of Earth as it appears right now.

____ 2. Rocks are actual pieces of Earth.

____ 3. Google Earth™ shows representations of the surface of planet Earth.

____ 4. Earth-imaging satellites can take a picture of the whole surface of one side of Earth.

Write a few sentences explaining how this article helped you clarify your thinking about Earth imaging.

Anticipation Guide

Before reading the article "Seeing Earth," check off the statements with which you agree.

____ 1. Images in Google Earth™ show live images of Earth as it appears right now.

____ 2. Rocks are actual pieces of Earth.

____ 3. Google Earth™ shows representations of the surface of planet Earth.

____ 4. Earth-imaging satellites can take a picture of the whole surface of one side of Earth.

After reading the article, check off the statements that were confirmed in the text.

____ 1. Images in Google Earth™ show live images of Earth as it appears right now.

____ 2. Rocks are actual pieces of Earth.

____ 3. Google Earth™ shows representations of the surface of planet Earth.

____ 4. Earth-imaging satellites can take a picture of the whole surface of one side of Earth.

Write a few sentences explaining how this article helped you clarify your thinking about Earth imaging.

FOSS Next Generation
© The Regents of the University of California
Can be duplicated for classroom or workshop use.
Earth History Course
Investigation 1: Earth Is Rock
No. 3—Notebook Master

Mile 20 Sketch

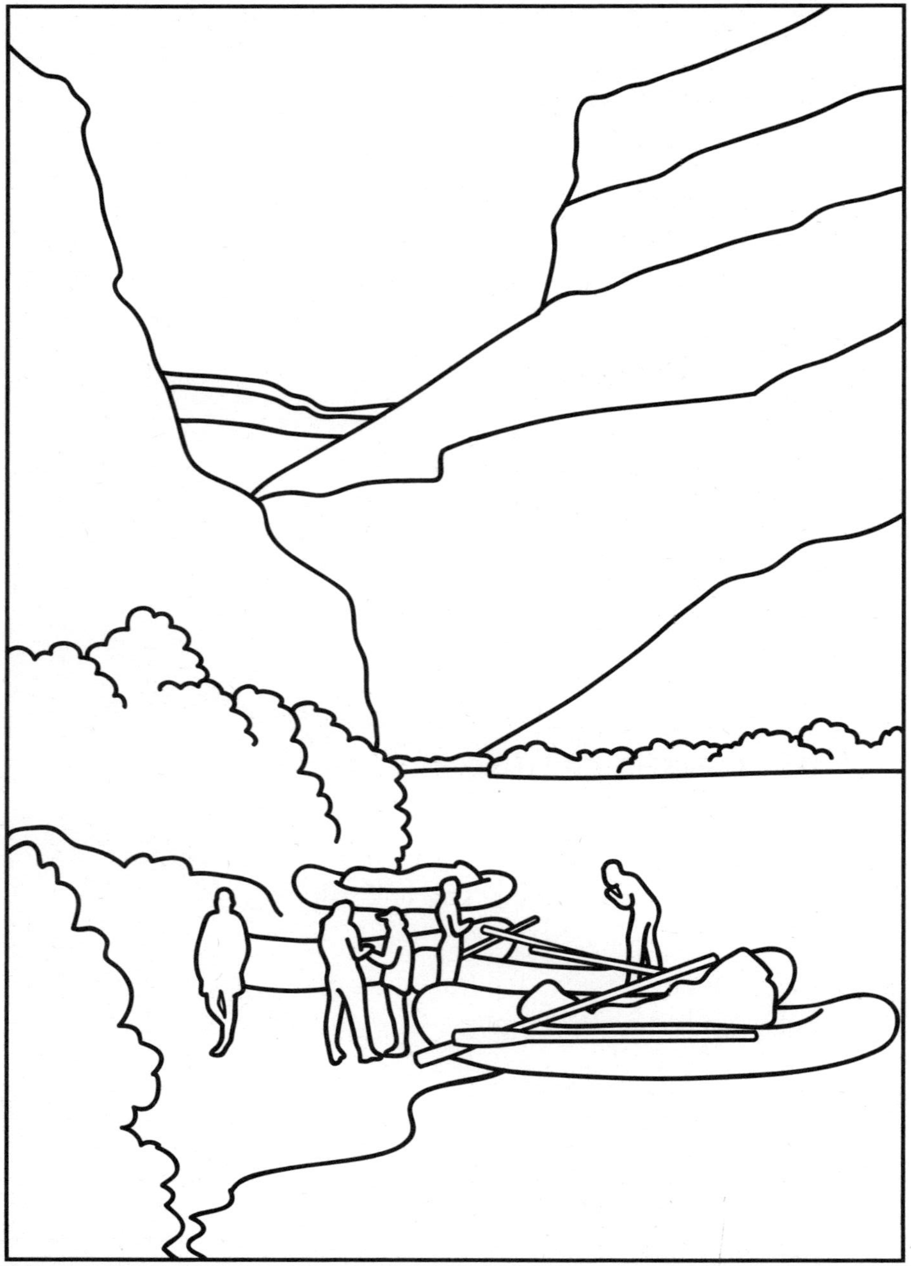

Mile 20 Sketch

FOSS Next Generation
© The Regents of the University of California
Can be duplicated for classroom or workshop use.
Earth History Course
Investigation 1: Earth Is Rock
No. 4—Notebook Master

Mile 52 Sketch

Mile 52 Sketch

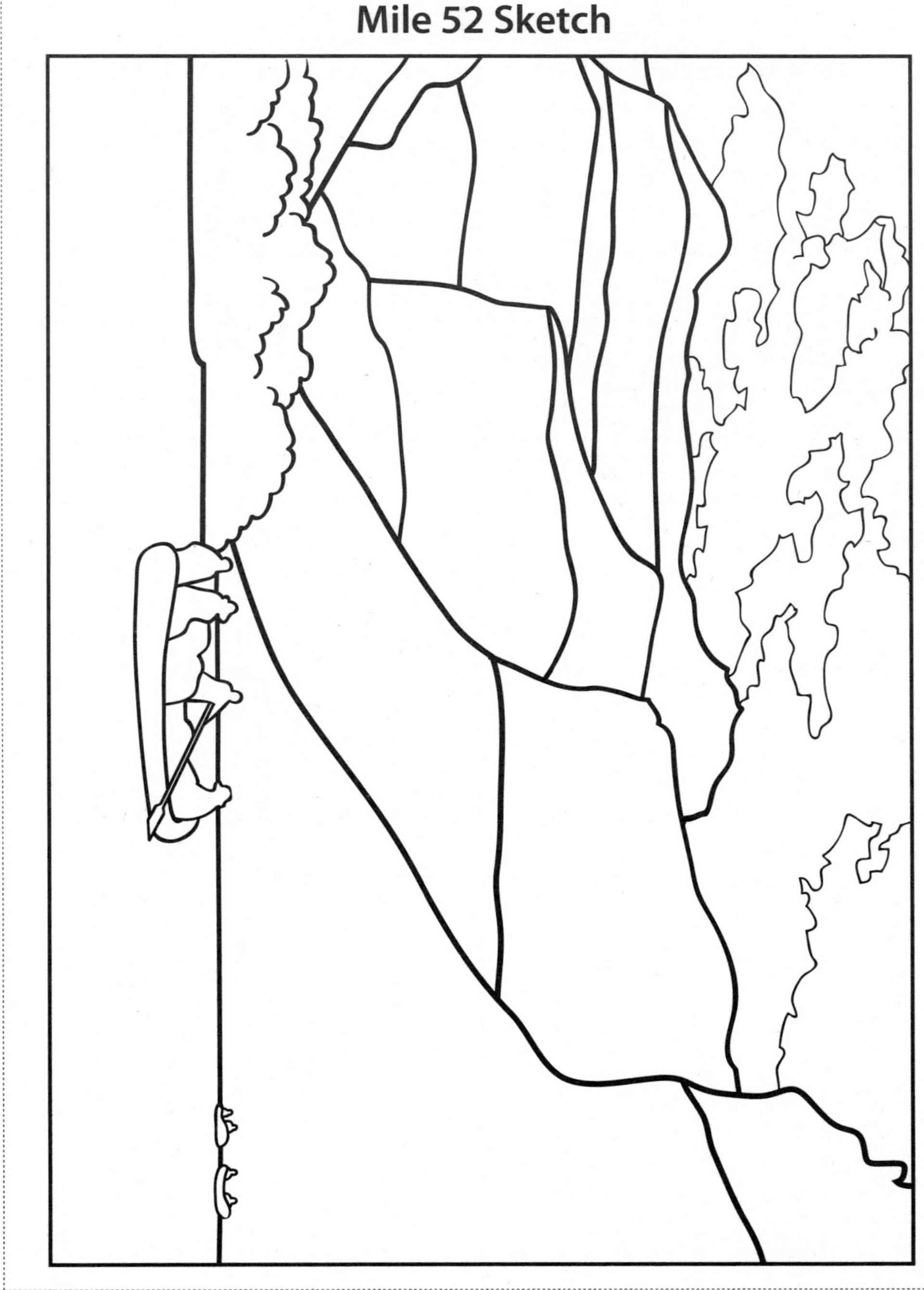

FOSS Next Generation
© The Regents of the University of California
Can be duplicated for classroom or workshop use.

Earth History Course
Investigation 1: Earth Is Rock
No. 5—Notebook Master

Mile 20 Rock Observations

WARNING — This set contains chemicals that may be harmful if misused. Read cautions on individual containers carefully. Not to be used by children except under adult supervision.

MILE 20 ROCK OBSERVATIONS

Rock number	Color	Texture	Observations from photos	Other

Mile 20 Rock Observations

WARNING — This set contains chemicals that may be harmful if misused. Read cautions on individual containers carefully. Not to be used by children except under adult supervision.

MILE 20 ROCK OBSERVATIONS

Rock number	Color	Texture	Observations from photos	Other

Earth History Course
Investigation 1: Earth Is Rock
No. 6—Notebook Master

FOSS Next Generation
© The Regents of the University of California
Can be duplicated for classroom or workshop use.

Mile 52 Rock Observations

WARNING — This set contains chemicals that may be harmful if misused. Read cautions on individual containers carefully. Not to be used by children except under adult supervision.

MILE 52 ROCK OBSERVATIONS				
Rock number	Color	Texture	Observations from photos	Other

Mile 52 Rock Observations

WARNING — This set contains chemicals that may be harmful if misused. Read cautions on individual containers carefully. Not to be used by children except under adult supervision.

MILE 52 ROCK OBSERVATIONS				
Rock number	Color	Texture	Observations from photos	Other

FOSS Next Generation
© The Regents of the University of California
Can be duplicated for classroom or workshop use.

Earth History Course
Investigation 1: Earth Is Rock
No. 7—Notebook Master

Grand Canyon Rocks

Mile 52	
Rock	**Rock-layer name**
9	Limestone
8	Sandstone
7	Shale
6	Sandstone
5	Limestone
4	Limestone
Colorado River	
Elevation of river: 853 meters	

Mile 20	
Rock	**Rock-layer name**
10	Limestone
9	Limestone
8	Sandstone
7	Shale
6	Sandstone
Colorado River	
Elevation of river: 892 meters	

Grand Canyon Rocks

Mile 52	
Rock	**Rock-layer name**
9	Limestone
8	Sandstone
7	Shale
6	Sandstone
5	Limestone
4	Limestone
Colorado River	
Elevation of river: 853 meters	

Mile 20	
Rock	**Rock-layer name**
10	Limestone
9	Limestone
8	Sandstone
7	Shale
6	Sandstone
Colorado River	
Elevation of river: 892 meters	

Earth History Course
Investigation 1: Earth Is Rock
No. 8—Notebook Master

FOSS Next Generation
© The Regents of the University of California
Can be duplicated for classroom or workshop use.

Correlation Questions

Use your "Grand Canyon Rock Lineup" and *FOSS Science Resources* to answer these questions.

1. How far apart are the sites at Mile 20 and Mile 52?
2. What is the elevation of the river at Mile 20?
3. What is the elevation of the river at Mile 52?
4. Which way is the Colorado River flowing, from Mile 20 to Mile 52 or from Mile 52 to Mile 20? How do you know?
5. Which rock layer is at river level at Mile 20?
6. Which rock layer is at river level at Mile 52?
7. Why do these two sites have different rock layers exposed at river level?
8. Suppose you were in a boat on the river at Mile 20 and you could drill down into the rock under the river. What kind of rock would you expect to find? Why?
9. Suppose you stopped at Mile 36 along the Colorado River in the Grand Canyon. Which rock layer would you expect to see at river level? Why?

Correlation Questions

Use your "Grand Canyon Rock Lineup" and *FOSS Science Resources* to answer these questions.

1. How far apart are the sites at Mile 20 and Mile 52?
2. What is the elevation of the river at Mile 20?
3. What is the elevation of the river at Mile 52?
4. Which way is the Colorado River flowing, from Mile 20 to Mile 52 or from Mile 52 to Mile 20? How do you know?
5. Which rock layer is at river level at Mile 20?
6. Which rock layer is at river level at Mile 52?
7. Why do these two sites have different rock layers exposed at river level?
8. Suppose you were in a boat on the river at Mile 20 and you could drill down into the rock under the river. What kind of rock would you expect to find? Why?
9. Suppose you stopped at Mile 36 along the Colorado River in the Grand Canyon. Which rock layer would you expect to see at river level? Why?

FOSS Next Generation
© The Regents of the University of California
Can be duplicated for classroom or workshop use.
Earth History Course
Investigation 1: Earth Is Rock
No. 9—Notebook Master

Stream-Table Map

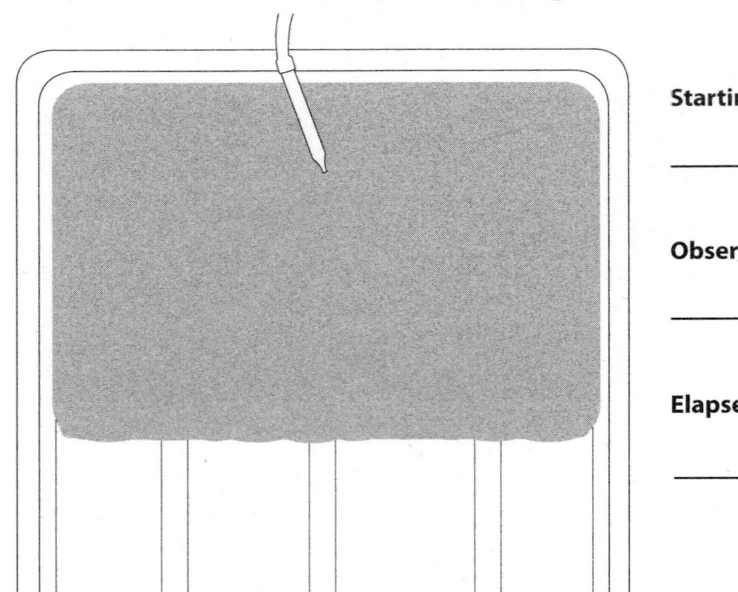

Starting time ______________

Observation time ______________

Elapsed time ______________

1. Observe where different materials are being deposited (that is, sand, clay), and label them on the map.
2. Label on the map where the water is flowing fastest and where it is flowing slowest.
3. Label on the map any landforms that have been created.

Stream-Table Map

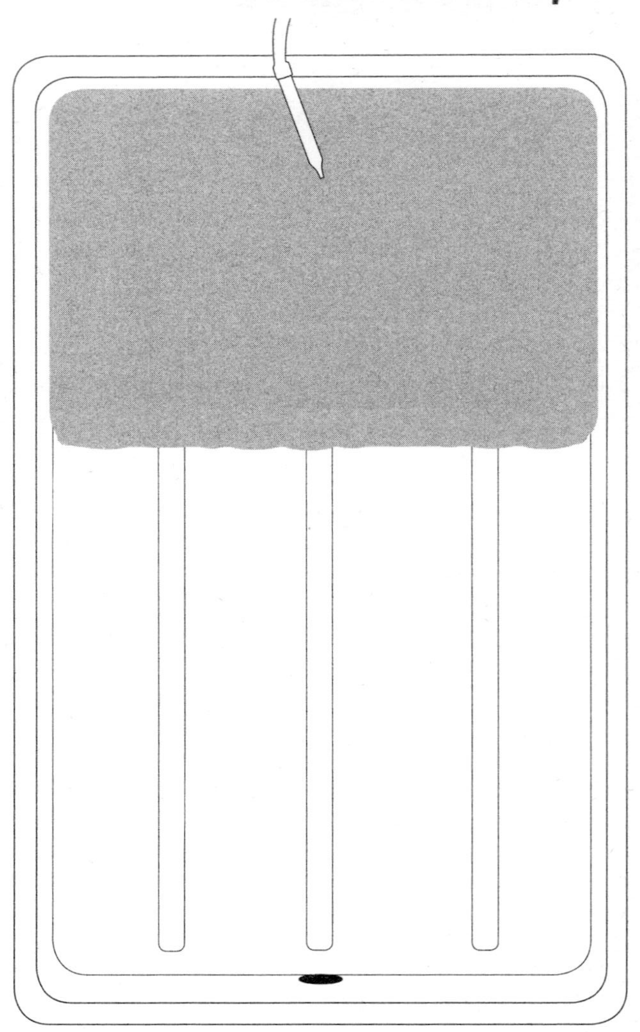

Starting time ______________

Observation time ______________

Elapsed time ______________

1. Observe where different materials are being deposited (that is, sand, clay), and label them on the map.
2. Label on the map where the water is flowing fastest and where it is flowing slowest.
3. Label on the map any landforms that have been created.

FOSS Next Generation
© The Regents of the University of California
Can be duplicated for classroom or workshop use.
Earth History Course
Investigation 2: Weathering and Erosion
No. 10—Notebook Master

Stream-Table Questions

Refer to your *Stream-Table Map* as you answer these questions.

1. Watch a grain of sand as it moves along. Describe its motion.

2. Where are the large particles deposited? The small particles?

3. Is a delta forming? Where? Why is it forming there?

4. What color is the water flowing into the basin?

5. Consider the Grand Canyon, and refer to the Colorado Plateau map in *FOSS Science Resources*. Where is the material eroded by the Colorado River deposited today?

6. Where do you think the material that was eroded by the Colorado River was deposited in the past?

7. Which do you think came first, the Colorado Plateau, the Colorado River, or the Grand Canyon? Support your answer with evidence.

Stream-Table Questions

Refer to your *Stream-Table Map* as you answer these questions.

1. Watch a grain of sand as it moves along. Describe its motion.

2. Where are the large particles deposited? The small particles?

3. Is a delta forming? Where? Why is it forming there?

4. What color is the water flowing into the basin?

5. Consider the Grand Canyon, and refer to the Colorado Plateau map in *FOSS Science Resources*. Where is the material eroded by the Colorado River deposited today?

6. Where do you think the material that was eroded by the Colorado River was deposited in the past?

7. Which do you think came first, the Colorado Plateau, the Colorado River, or the Grand Canyon? Support your answer with evidence.

FOSS Next Generation
© The Regents of the University of California
Can be duplicated for classroom or workshop use.
Earth History Course
Investigation 2: Weathering and Erosion
No. 11—Notebook Master

Stream-Table Videos

Part 1: Flood

1. Describe in detail the motion of earth materials you see in a flood condition.

Part 2: Different earth materials

Homogeneous = mixture of sand and clay just like the material used in class

Heterogeneous = mixture of sand and clay on top and bottom; layer of red clay in the middle

2. In which stream table is the earth material eroding faster and deeper?

3. What is happening to the top sand/clay level in the heterogeneous materials?

4. What is happening to the layer of red clay?

5. The bottom layer was made out of the same material as the top layer. Why didn't it erode as quickly?

Stream-Table Videos

Part 1: Flood

1. Describe in detail the motion of earth materials you see in a flood condition.

Part 2: Different earth materials

Homogeneous = mixture of sand and clay just like the material used in class

Heterogeneous = mixture of sand and clay on top and bottom; layer of red clay in the middle

2. In which stream table is the earth material eroding faster and deeper?

3. What is happening to the top sand/clay level in the heterogeneous materials?

4. What is happening to the layer of red clay?

5. The bottom layer was made out of the same material as the top layer. Why didn't it erode as quickly?

FOSS Next Generation
© The Regents of the University of California
Can be duplicated for classroom or workshop use.

Earth History Course
Investigation 2: Weathering and Erosion
No. 12—Notebook Master

Interesting rock formations like arches and pillars in Utah's Bryce Canyon were created by the natural forces of weathering.

Weathering and Erosion

What do sheer cliffs, balancing rocks, massive caves, and giant sand dunes have in common? They result from the processes of weathering and erosion. The same processes form and take away the soils we depend on to grow our food.

How can **weathering** and erosion be all that? The investigations you have done so far give some clues about these processes. Remember, in **physical weathering**, rock breaks down into smaller pieces. The smaller pieces are called sediment. Erosion transports sediment to a **basin** by water, wind, or ice.

Interesting rock formations like arches and pillars in Utah's Bryce Canyon were created by the natural forces of weathering.

Weathering and Erosion

What do sheer cliffs, balancing rocks, massive caves, and giant sand dunes have in common? They result from the processes of weathering and erosion. The same processes form and take away the soils we depend on to grow our food.

How can **weathering** and erosion be all that? The investigations you have done so far give some clues about these processes. Remember, in **physical weathering**, rock breaks down into smaller pieces. The smaller pieces are called sediment. Erosion transports sediment to a **basin** by water, wind, or ice.

20

FOSS Next Generation
© The Regents of the University of California
Can be duplicated for classroom or workshop use.
Earth History Course
Investigation 2: Weathering and Erosion
No. 13—Notebook Master

The abrasion, or scraping, of blowing sand helped carve and smooth this sandstone canyon.

Physical Weathering and Round Rocks

Physical weathering occurs when large rocks break into smaller rocks of the same kind. When a rock, like granite, is broken, it may break into small pieces of the **minerals** that make it up, such as quartz and feldspar. But they are the same minerals that were in the original granite.

The sharp edges and corners of broken rock pieces wear away as they hit other rocks. This reshaping occurs naturally when rocks are hit by windblown sand or rock particles in moving water. The name for this type of physical weathering is **abrasion**. Abrasion also happens when falling rocks hit other rocks, breaking them apart.

When you observe beach sand or sand in a riverbed, you can see smooth, polished sand grains. Waves and flowing water rolled these sand grains around, causing them to hit each other. You observed sand particles in the stream table bouncing and hitting other grains of sand as they moved along. The water carries rocks that bump off the rough edges on other rocks. The farther the sand grains are carried by water, the smoother they get.

The abrasion, or scraping, of blowing sand helped carve and smooth this sandstone canyon.

Physical Weathering and Round Rocks

Physical weathering occurs when large rocks break into smaller rocks of the same kind. When a rock, like granite, is broken, it may break into small pieces of the minerals that make it up, such as quartz and feldspar. But they are the same minerals that were in the original granite.

The sharp edges and corners of broken rock pieces wear away as they hit other rocks. This reshaping occurs naturally when rocks are hit by windblown sand or rock particles in moving water. The name for this type of physical weathering is abrasion. Abrasion also happens when falling rocks hit other rocks, breaking them apart.

When you observe beach sand or sand in a riverbed, you can see smooth, polished sand grains. Waves and flowing water rolled these sand grains around, causing them to hit each other. You observed sand particles in the stream table bouncing and hitting other grains of sand as they moved along. The water carries rocks that bump off the rough edges on other rocks. The farther the sand grains are carried by water, the smoother they get.

Investigation 2: Weathering and Erosion 21

FOSS Next Generation
© The Regents of the University of California
Can be duplicated for classroom or workshop use.

Earth History Course
Investigation 2: Weathering and Erosion
No. 14—Notebook Master

No water is involved when wind transports sediment. The sand grains bang into each other, creating a frosted, dull rounded surface. Beach sand, river sand, and dune sand are all similar in at least one way—the farther the weathered rocks travel and the more they get banged around, the smaller they become.

Ice Wedging and Rock Falls

When ice freezes, it expands with great force. You saw what happened when water in a jar froze and expanded. The force shattered the jar! Ice expansion naturally causes physical weathering when water gets into tiny cracks in a rock. At night, temperatures fall, and the water freezes, expands, and presses against the surrounding rock. The crack gets bigger. During the day as the temperature rises, the water thaws and seeps farther down into the crack. Night comes, and the water freezes again. With repeated freezing and thawing, the crack becomes larger. Eventually, pieces of the rock break off. Ice wedging can break rocks off the side of a cliff. These **rock falls** make piles of jagged rocks, called **talus**, at the base of cliffs. These rocks fell because a tiny crack kept growing until the piece broke off. You may have seen the results of ice wedging in your neighborhood. Ice damages concrete sidewalks and curbs, and wedges flakes from bricks.

Plant roots also cause weathering, by growing into cracks. The roots expand as the plant grows, breaking the rocks apart. You may have seen tree roots that lifted and broke a sidewalk or even cracked the foundation of a house.

The slope of loose rock, or talus, at the base of the cliff is evidence of rock falls.

Tree roots cause physical weathering when they grow into cracks in a rock. Growing roots can split huge boulders.

No water is involved when wind transports sediment. The sand grains bang into each other, creating a frosted, dull rounded surface. Beach sand, river sand, and dune sand are all similar in at least one way—the farther the weathered rocks travel and the more they get banged around, the smaller they become.

Ice Wedging and Rock Falls

When ice freezes, it expands with great force. You saw what happened when water in a jar froze and expanded. The force shattered the jar! Ice expansion naturally causes physical weathering when water gets into tiny cracks in a rock. At night, temperatures fall, and the water freezes, expands, and presses against the surrounding rock. The crack gets bigger. During the day as the temperature rises, the water thaws and seeps farther down into the crack. Night comes, and the water freezes again. With repeated freezing and thawing, the crack becomes larger. Eventually, pieces of the rock break off. Ice wedging can break rocks off the side of a cliff. These **rock falls** make piles of jagged rocks, called **talus**, at the base of cliffs. These rocks fell because a tiny crack kept growing until the piece broke off. You may have seen the results of ice wedging in your neighborhood. Ice damages concrete sidewalks and curbs, and wedges flakes from bricks.

Plant roots also cause weathering, by growing into cracks. The roots expand as the plant grows, breaking the rocks apart. You may have seen tree roots that lifted and broke a sidewalk or even cracked the foundation of a house.

The slope of loose rock, or talus, at the base of the cliff is evidence of rock falls.

Tree roots cause physical weathering when they grow into cracks in a rock. Growing roots can split huge boulders.

22

FOSS Next Generation
© The Regents of the University of California
Can be duplicated for classroom or workshop use.

Earth History Course
Investigation 2: Weathering and Erosion
No. 15—Notebook Master

Vast karst landscapes are found in the humid tropics of Southeast Asia. The towers, pinnacles, and cones are caused by chemical weathering.

Chemical Weathering

Physical weathering is not the only way to break down rocks. Remember how the limestone fizzed when you put a drop of acid on it? During that **chemical reaction**, acid dissolved a tiny amount of rock. This same process takes place naturally. Tiny amounts of carbon dioxide in the air dissolve in falling raindrops. The solution is a very weak acid called carbonic acid. This acid is too weak to make limestone fizz. But each slightly acidic raindrop dissolves a few **molecules** of limestone. Over thousands of years, limestone will slowly wear away because of **chemical weathering**.

Decaying plant material, lichens, and plant roots also produce carbon dioxide and other weak acids. These acids dissolve in rainwater as it moves through **soil** and into cracks in rock. Limestone caves such as Mammoth Cave in Kentucky formed over millions of years as chemical weathering dissolved limestone. This weathering creates landforms called **karst topography**. Sinkholes and caves are karst landforms and are found in many areas, including Kentucky and Florida. Dramatic limestone pinnacles found in Asia are also examples of karst topography. Because acids need moisture to form, karst topography is found in humid environments. In dry locations, limestone tends to form steep cliffs, like the Redwall Limestone in the Grand Canyon.

Vast karst landscapes are found in the humid tropics of Southeast Asia. The towers, pinnacles, and cones are caused by chemical weathering.

Chemical Weathering

Physical weathering is not the only way to break down rocks. Remember how the limestone fizzed when you put a drop of acid on it? During that **chemical reaction**, acid dissolved a tiny amount of rock. This same process takes place naturally. Tiny amounts of carbon dioxide in the air dissolve in falling raindrops. The solution is a very weak acid called carbonic acid. This acid is too weak to make limestone fizz. But each slightly acidic raindrop dissolves a few **molecules** of limestone. Over thousands of years, limestone will slowly wear away because of **chemical weathering**.

Decaying plant material, lichens, and plant roots also produce carbon dioxide and other weak acids. These acids dissolve in rainwater as it moves through **soil** and into cracks in rock. Limestone caves such as Mammoth Cave in Kentucky formed over millions of years as chemical weathering dissolved limestone. This weathering creates landforms called **karst topography**. Sinkholes and caves are karst landforms and are found in many areas, including Kentucky and Florida. Dramatic limestone pinnacles found in Asia are also examples of karst topography. Because acids need moisture to form, karst topography is found in humid environments. In dry locations, limestone tends to form steep cliffs, like the Redwall Limestone in the Grand Canyon.

Investigation 2: Weathering and Erosion 23

FOSS Next Generation
© The Regents of the University of California
Can be duplicated for classroom or workshop use.

Earth History Course
Investigation 2: Weathering and Erosion
No. 16—Notebook Master

Weak acids also weather granite, though much more slowly than limestone. Granite is made of several common minerals, mainly quartz, feldspar, and hornblende. Quartz is very resistant to chemical weathering. Feldspar is easier to break down. Acid slowly weathers feldspar into clay particles. Without feldspar to hold the granite together, the quartz **crystals** fall out and become sand. The toughness of quartz is the reason so much sand is mostly quartz.

Differential Erosion

You saw **differential erosion** in action in the stream tables that had a layer of clay between two layers of sand. The water easily eroded away the top layer of sand. The clay layer resisted erosion. As long as the clay layer was solid, it protected the bottom layer of sand. This is called differential erosion, because the layers erode at different rates.

This large mushroom rock near Lees Ferry on the Colorado River in Arizona is the result of differential erosion.
Warning: Rock structures of this kind can be hazardous. Be careful around them.

Once hidden below Earth's surface, Devils Tower now looms above the Wyoming prairie. Erosion has worn away the soft sandstone and shale that once overlaid this plug of hard volcanic rock.

Weak acids also weather granite, though much more slowly than limestone. Granite is made of several common minerals, mainly quartz, feldspar, and hornblende. Quartz is very resistant to chemical weathering. Feldspar is easier to break down. Acid slowly weathers feldspar into clay particles. Without feldspar to hold the granite together, the quartz **crystals** fall out and become sand. The toughness of quartz is the reason so much sand is mostly quartz.

Differential Erosion

You saw **differential erosion** in action in the stream tables that had a layer of clay between two layers of sand. The water easily eroded away the top layer of sand. The clay layer resisted erosion. As long as the clay layer was solid, it protected the bottom layer of sand. This is called differential erosion, because the layers erode at different rates.

This large mushroom rock near Lees Ferry on the Colorado River in Arizona is the result of differential erosion.
Warning: Rock structures of this kind can be hazardous. Be careful around them.

Once hidden below Earth's surface, Devils Tower now looms above the Wyoming prairie. Erosion has worn away the soft sandstone and shale that once overlaid this plug of hard volcanic rock.

24

FOSS Next Generation
© The Regents of the University of California
Can be duplicated for classroom or workshop use.
Earth History Course
Investigation 2: Weathering and Erosion
No. 17—Notebook Master

The Niagara River tumbles over a hard limestone layer to form a waterfall 50 m high. Millions of gallons of water plunge downward every minute.

Differential erosion happens any time soft rock is eroded away, leaving harder rock behind. Much of the scenery in the Grand Canyon is due to differential erosion. Devils Tower in Wyoming consists of hard rock that was once surrounded by softer rock. Over the past 1 to 2 million years, the softer rock weathered and eroded away. The column of hard rock still stands.

Niagara Falls on the New York–Canada border is another example of differential erosion. The water going over the falls erodes the edge of the thick, soft **shale** layer under a hard limestone layer. This undercuts the limestone, causing it to give way. The photo above of American Falls, a part of Niagara Falls, shows huge limestone boulders that have fallen. For the past 10,000 years, the falls have moved upstream, eroding the rock at the plunging edge of the fall, at an average rate of about 1 meter (m) a year.

Wind Erosion and Rain Forests

How could erosion in Africa help the Amazon rain forests, all the way across the Atlantic Ocean? Windstorms on the Sahara Desert carry dust high into the **atmosphere**. High-altitude winds carry the dust west across the Atlantic Ocean. Sometimes it reaches the rain forests of South and Central America. The soil in rain forests is normally poor in nutrients, and the dust from Africa provides many of the nutrients that the rain forest plants need to survive.

The Niagara River tumbles over a hard limestone layer to form a waterfall 50 m high. Millions of gallons of water plunge downward every minute.

Differential erosion happens any time soft rock is eroded away, leaving harder rock behind. Much of the scenery in the Grand Canyon is due to differential erosion. Devils Tower in Wyoming consists of hard rock that was once surrounded by softer rock. Over the past 1 to 2 million years, the softer rock weathered and eroded away. The column of hard rock still stands.

Niagara Falls on the New York–Canada border is another example of differential erosion. The water going over the falls erodes the edge of the thick, soft **shale** layer under a hard limestone layer. This undercuts the limestone, causing it to give way. The photo above of American Falls, a part of Niagara Falls, shows huge limestone boulders that have fallen. For the past 10,000 years, the falls have moved upstream, eroding the rock at the plunging edge of the fall, at an average rate of about 1 meter (m) a year.

Wind Erosion and Rain Forests

How could erosion in Africa help the Amazon rain forests, all the way across the Atlantic Ocean? Windstorms on the Sahara Desert carry dust high into the **atmosphere**. High-altitude winds carry the dust west across the Atlantic Ocean. Sometimes it reaches the rain forests of South and Central America. The soil in rain forests is normally poor in nutrients, and the dust from Africa provides many of the nutrients that the rain forest plants need to survive.

Investigation 2: Weathering and Erosion 25

FOSS Next Generation
© The Regents of the University of California
Can be duplicated for classroom or workshop use.

Earth History Course
Investigation 2: Weathering and Erosion
No. 18—Notebook Master

Dust storms can carry silt and other fine particles long distances, even across the ocean.

There is also a negative side to all this dust. Some of the dust settles on coral reefs in the sea. The dust carries bacteria and fungi that can kill or weaken the coral. When the dust settles over populated areas, it can trigger asthma and other respiratory diseases. Different places in the world are affected by different dust sources. Dust from China sometimes reaches people living in California!

Weathering and erosion created the Grand Canyon and many spectacular landforms around the world. The spires and **hoodoos** of Bryce Canyon in Utah, the rugged Badlands in South Dakota, and the rounded Blue Ridge Mountains extending from West Virginia to Pennsylvania are all products of weathering and erosion. Mammoth Cave in Kentucky, the world's longest cave system, was created by weathering and erosion. All these wonders were once solid rock.

Weathering and erosion produce sediments that can form new rocks, create soil, change landforms, and affect air quality. Think about that the next time you feel smooth, rounded sand between your toes at the beach, see a crack in a sidewalk, or eat a carrot!

Think Questions

1. **Choose one of the photos of rock formations in this article. Make a sketch of the formation. Label the layers that you think might be hard rock and those that might be softer. What is your evidence?**
2. **Describe the processes you think might have produced the mushroom-shaped rock in the photo.**
3. **Think about your community. Give at least one example of where you have seen these processes.**
   - **Weathering**
   - **Erosion**
   - **Differential erosion**

Dust storms can carry silt and other fine particles long distances, even across the ocean.

There is also a negative side to all this dust. Some of the dust settles on coral reefs in the sea. The dust carries bacteria and fungi that can kill or weaken the coral. When the dust settles over populated areas, it can trigger asthma and other respiratory diseases. Different places in the world are affected by different dust sources. Dust from China sometimes reaches people living in California!

Weathering and erosion created the Grand Canyon and many spectacular landforms around the world. The spires and **hoodoos** of Bryce Canyon in Utah, the rugged Badlands in South Dakota, and the rounded Blue Ridge Mountains extending from West Virginia to Pennsylvania are all products of weathering and erosion. Mammoth Cave in Kentucky, the world's longest cave system, was created by weathering and erosion. All these wonders were once solid rock.

Weathering and erosion produce sediments that can form new rocks, create soil, change landforms, and affect air quality. Think about that the next time you feel smooth, rounded sand between your toes at the beach, see a crack in a sidewalk, or eat a carrot!

Think Questions

1. **Choose one of the photos of rock formations in this article. Make a sketch of the formation. Label the layers that you think might be hard rock and those that might be softer. What is your evidence?**
2. **Describe the processes you think might have produced the mushroom-shaped rock in the photo.**
3. **Think about your community. Give at least one example of where you have seen these processes.**
   - **Weathering**
   - **Erosion**
   - **Differential erosion**

26

FOSS Next Generation
© The Regents of the University of California
Can be duplicated for classroom or workshop use.
Earth History Course
Investigation 2: Weathering and Erosion
No. 19—Notebook Master

Response Sheet—Investigation 2

A group of students were collecting soil samples near a river. One student noted there was a lot of sand in one of the samples.

He said, "All this sand must have washed up from the beach where the river enters the ocean."

Another student said, "I think it came from the rocks in the mountains. But there aren't any mountains near us. I don't understand how the sand got there."

What would you tell the students to help them solve where the sand came from?

Response Sheet—Investigation 2

A group of students were collecting soil samples near a river. One student noted there was a lot of sand in one of the samples.

He said, "All this sand must have washed up from the beach where the river enters the ocean."

Another student said, "I think it came from the rocks in the mountains. But there aren't any mountains near us. I don't understand how the sand got there."

What would you tell the students to help them solve where the sand came from?

FOSS Next Generation
© The Regents of the University of California
Can be duplicated for classroom or workshop use.
Earth History Course
Investigation 2: Weathering and Erosion
No. 20—Notebook Master

Seawater Investigation

WARNING — This set contains chemicals that may be harmful if misused. Read cautions on individual containers carefully. Not to be used by children except under adult supervision.

Materials

1 Plastic cup with lid
60 mL Limewater (calcium hydroxide solution)
4 Straws with hole punched in side
4 Safety goggles

Instructions

1. Work with your group. Measure 60 mL of limewater into a cup. Limewater is a $Ca(OH)_2$ solution.
2. Place the lid on the cup.
3. Use the table below to record observations before bubbling.
4. Take turns poking your straw through the hole in the lid and gently blowing air into the limewater. Continue taking turns for 2 or 3 minutes. **SAFETY NOTE: Don't suck up the limewater.** Make sure you don't blow so hard that the water splatters. If you get some limewater on your hands, rinse them with clear water.
5. Record observations after bubbling.
6. Let the cup stand for 5 minutes and then record your observations.

Seawater Observations

Observations of $Ca(OH)_2$ cup *before* bubbling	Observations of $Ca(OH)_2$ cup *after* bubbling	Observations of $Ca(OH)_2$ cup *after* standing for 5 minutes

Analysis: What do you think the limewater reaction has to do with limestone formation?

Seawater Investigation

WARNING — This set contains chemicals that may be harmful if misused. Read cautions on individual containers carefully. Not to be used by children except under adult supervision.

Materials

1 Plastic cup with lid
60 mL Limewater (calcium hydroxide solution)
4 Straws with hole punched in side
4 Safety goggles

Instructions

1. Work with your group. Measure 60 mL of limewater into a cup. Limewater is a $Ca(OH)_2$ solution.
2. Place the lid on the cup.
3. Use the table below to record observations before bubbling.
4. Take turns poking your straw through the hole in the lid and gently blowing air into the limewater. Continue taking turns for 2 or 3 minutes. **SAFETY NOTE: Don't suck up the limewater.** Make sure you don't blow so hard that the water splatters. If you get some limewater on your hands, rinse them with clear water.
5. Record observations after bubbling.
6. Let the cup stand for 5 minutes and then record your observations.

Seawater Observations

Observations of $Ca(OH)_2$ cup *before* bubbling	Observations of $Ca(OH)_2$ cup *after* bubbling	Observations of $Ca(OH)_2$ cup *after* standing for 5 minutes

Analysis: What do you think the limewater reaction has to do with limestone formation?

FOSS Next Generation
© The Regents of the University of California
Can be duplicated for classroom or workshop use.
Earth History Course
Investigation 3: Deposition
No. 21—Notebook Master

Basin Questions

Use the correlation of Grand Canyon rocks from Investigation 1 to help answer these questions.

1. Which Grand Canyon rock layer is the oldest that we have observed so far?

2. How do you know it is the oldest?

3. Which layer in the Grand Canyon is the youngest that we have observed so far?

4. How do you know it is the youngest?

5. What do you think is below the oldest layer?

Basin Questions

Use the correlation of Grand Canyon rocks from Investigation 1 to help answer these questions.

1. Which Grand Canyon rock layer is the oldest that we have observed so far?

2. How do you know it is the oldest?

3. Which layer in the Grand Canyon is the youngest that we have observed so far?

4. How do you know it is the youngest?

5. What do you think is below the oldest layer?

FOSS Next Generation
© The Regents of the University of California
Can be duplicated for classroom or workshop use.
Earth History Course
Investigation 3: Deposition
No. 22—Notebook Master

Response Sheet—Investigation 4

A student's little brother said to her,

"The fossils in the Grand Canyon are dead bodies of animals that washed up onto the rocks as the river passed by."

If you were the student, what would you say to your brother?

Response Sheet—Investigation 4

A student's little brother said to her,

"The fossils in the Grand Canyon are dead bodies of animals that washed up onto the rocks as the river passed by."

If you were the student, what would you say to your brother?

FOSS Next Generation
© The Regents of the University of California
Can be duplicated for classroom or workshop use.
Earth History Course
Investigation 4: Fossils and Past Environments
No. 23—Notebook Master

Grand Canyon Environments

Rock layer	Rock evidence	Fossil evidence
Kaibab Formation	Mostly limestone containing some grains of sand.	Sponges, corals, brachiopods, clams, and snails.
Toroweap Formation	Mostly limestone with some sandstone and siltstone layers.	Sponges, corals, brachiopods, clams, snails, and crinoids.
Coconino Sandstone	Sandstone with broken rock fragments. Well-sorted sand grains. Large crossbeds.	Reptile and insect tracks.
Hermit Shale	Shaley siltstone in many areas. Raindrop imprints and mud cracks.	Plant fossils, including arid-climate ferns and conifers, insects, worm trails, reptile or amphibian tracks.
Supai Group	Red and tan sandstones, siltstones, and a few limestones.	Vertebrate tracks in the sandstone layers, some brachiopods in the limestone layers. Fossils few and far between.
Redwall Limestone	Thick gray limestone stained red from iron oxide (rust).	Brachiopods, corals, crinoids, and bryozoans common. Most fossils whole, but much limestone made of fragments of fossilized shells.
Muav Limestone	Shaley, yellowish-gray limestone.	Trilobites, brachiopods.

Grand Canyon Environments

Rock layer	Rock evidence	Fossil evidence
Kaibab Formation	Mostly limestone containing some grains of sand.	Sponges, corals, brachiopods, clams, and snails.
Toroweap Formation	Mostly limestone with some sandstone and siltstone layers.	Sponges, corals, brachiopods, clams, snails, and crinoids.
Coconino Sandstone	Sandstone with broken rock fragments. Well-sorted sand grains. Large crossbeds.	Reptile and insect tracks.
Hermit Shale	Shaley siltstone in many areas. Raindrop imprints and mud cracks.	Plant fossils, including arid-climate ferns and conifers, insects, worm trails, reptile or amphibian tracks.
Supai Group	Red and tan sandstones, siltstones, and a few limestones.	Vertebrate tracks in the sandstone layers, some brachiopods in the limestone layers. Fossils few and far between.
Redwall Limestone	Thick gray limestone stained red from iron oxide (rust).	Brachiopods, corals, crinoids, and bryozoans common. Most fossils whole, but much limestone made of fragments of fossilized shells.
Muav Limestone	Shaley, yellowish-gray limestone.	Trilobites, brachiopods.

FOSS Next Generation
© The Regents of the University of California
Can be duplicated for classroom or workshop use.
Earth History Course
Investigation 4: Fossils and Past Environments
No. 24—Notebook Master

Geologic Time Calculations

Era	Period	Age (years)	Distance on time line (mm)	Distance on time line (cm)
Cenozoic	(Today) Quaternary	0.00		
	Tertiary	2,600,000		
Mesozoic	Cretaceous	66,000,000		
	Jurassic	145,000,000		
	Triassic	201,000,000		
Paleozoic	Permian	252,000,000		
	Pennsylvanian	299,000,000		
	Mississippian	323,000,000		
	Devonian	359,000,000		
	Silurian	419,000,000		
	Ordovician	444,000,000		
	Cambrian	485,000,000		
Precambrian		541,000,000		
		4,600,000,000		

1 mm = 1,000,000 years

Geologic Time Calculations

Era	Period	Age (years)	Distance on time line (mm)	Distance on time line (cm)
Cenozoic	(Today) Quaternary	0.00		
	Tertiary	2,600,000		
Mesozoic	Cretaceous	66,000,000		
	Jurassic	145,000,000		
	Triassic	201,000,000		
Paleozoic	Permian	252,000,000		
	Pennsylvanian	299,000,000		
	Mississippian	323,000,000		
	Devonian	359,000,000		
	Silurian	419,000,000		
	Ordovician	444,000,000		
	Cambrian	485,000,000		
Precambrian		541,000,000		
		4,600,000,000		

1 mm = 1,000,000 years

FOSS Next Generation
© The Regents of the University of California
Can be duplicated for classroom or workshop use.
Earth History Course
Investigation 4: Fossils and Past Environments
No. 25—Notebook Master

Index-Fossil Correlations

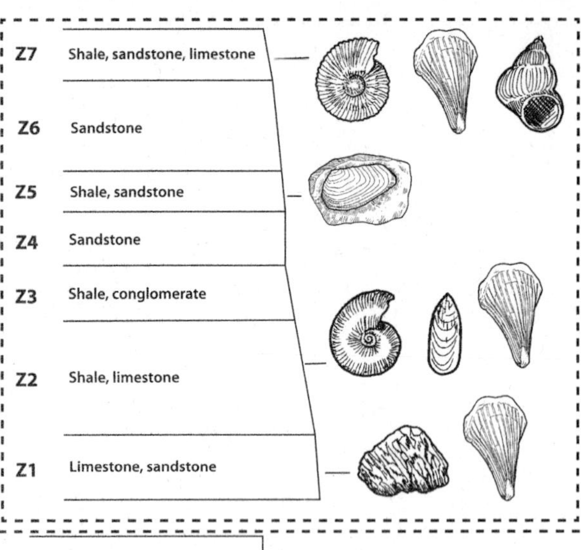

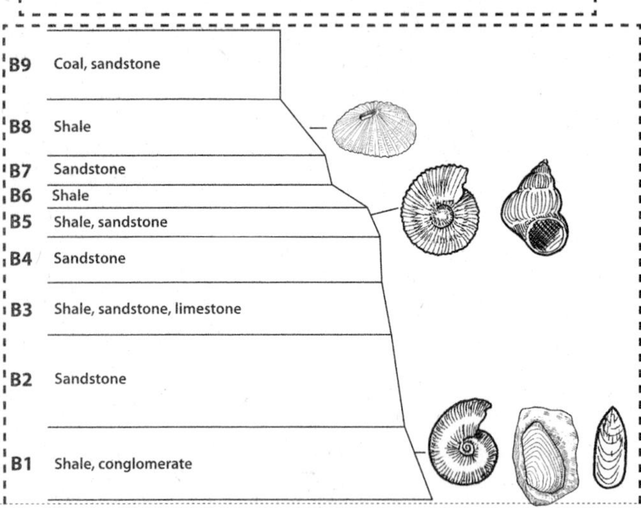

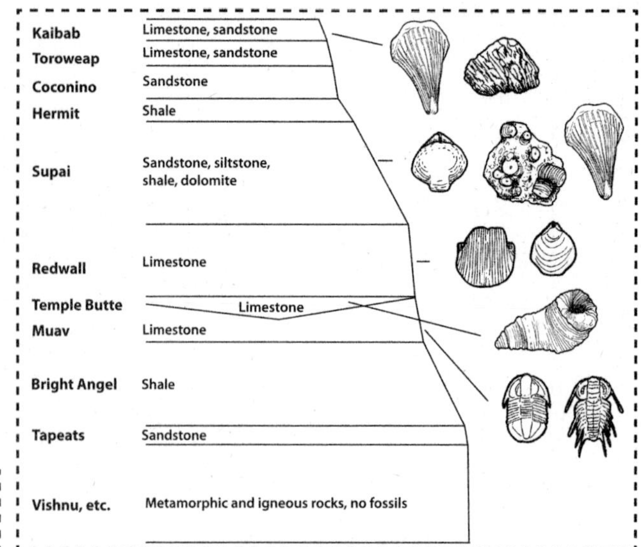

B = Bryce Canyon
Z = Zion National Park

Index-Fossil Correlations

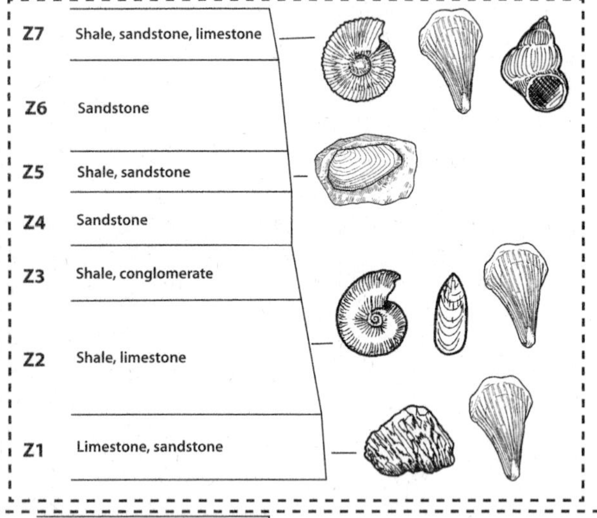

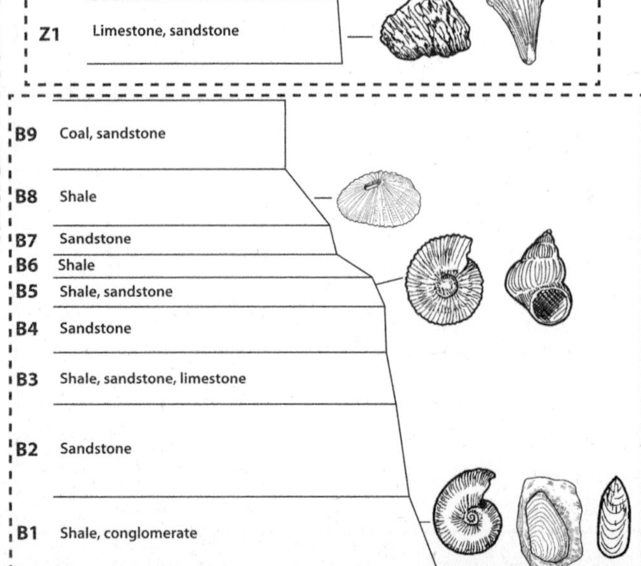

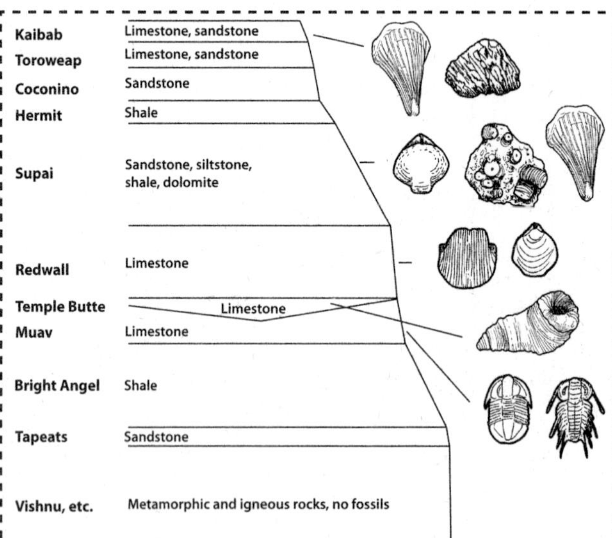

B = Bryce Canyon
Z = Zion National Park

Earth History Course
Investigation 4: Fossils and Past Environments
No. 26—Notebook Master
FOSS Next Generation
© The Regents of the University of California
Can be duplicated for classroom or workshop use.

Index-Fossil Correlation Questions

Answer these questions after you have identified and correlated the rock layers at the three parks.

1. Were any layers at all three canyons the same age?
2. Which rock layers contained the same index fossils at Zion and the Grand Canyon?
3. Which rock layers contained the same index fossils at Zion and Bryce?
4. Which rock layers contained the same index fossils at the Grand Canyon and Bryce?
5. Which canyon has the oldest rocks?
6. What was the age of the oldest rock layer?
7. Which canyon has the youngest rocks?
8. What was the age of the youngest rock layer?
9. Is rock layer B3 at Bryce older or younger than the Supai Group at the Grand Canyon? How do you know?
10. Is rock layer B2 at Bryce older or younger than rock layer Z1 at Zion? How do you know?

Index-Fossil Correlation Questions

Answer these questions after you have identified and correlated the rock layers at the three parks.

1. Were any layers at all three canyons the same age?
2. Which rock layers contained the same index fossils at Zion and the Grand Canyon?
3. Which rock layers contained the same index fossils at Zion and Bryce?
4. Which rock layers contained the same index fossils at the Grand Canyon and Bryce?
5. Which canyon has the oldest rocks?
6. What was the age of the oldest rock layer?
7. Which canyon has the youngest rocks?
8. What was the age of the youngest rock layer?
9. Is rock layer B3 at Bryce older or younger than the Supai Group at the Grand Canyon? How do you know?
10. Is rock layer B2 at Bryce older or younger than rock layer Z1 at Zion? How do you know?

FOSS Next Generation
© The Regents of the University of California
Can be duplicated for classroom or workshop use.
Earth History Course
Investigation 4: Fossils and Past Environments
No. 27—Notebook Master

Rocks over Time

Rock layer	Time of deposition (approximately)	Distance on time line	Period
Kaibab Formation	Ended 255,000,000 years ago Began 260,000,000 years ago	25.5 cm 26.0 cm	
Toroweap Formation	Ended 260,000,000 years ago Began 265,000,000 years ago	26.0 cm 26.5 cm	
Coconino Sandstone	Ended 265,000,000 years ago Began 270,000,000 years ago	26.5 cm 27.0 cm	
Hermit Shale	Ended 270,000,000 years ago Began 275,000,000 years ago	27.0 cm 27.5 cm	
Supai Group	Ended 275,000,000 years ago Began 325,000,000 years ago	27.5 cm 32.5 cm	
Redwall Limestone	Ended 325,000,000 years ago Began 360,000,000 years ago	32.5 cm 36.0 cm	
Temple Butte Limestone	Ended 370,000,000 years ago Began 375,000,000 years ago	37.0 cm 37.5 cm	
Muav Limestone	Ended 525,000,000 years ago Began 530,000,000 years ago	52.5 cm 53.0 cm	
Bright Angel Shale	Ended 530,000,000 years ago Began 540,000,000 years ago	53.0 cm 54.0 cm	
Tapeats Sandstone	Ended 540,000,000 years ago Began 545,000,000 years ago	54.0 cm 54.5 cm	

1 cm = 10,000,000 years

Rocks over Time

Rock layer	Time of deposition (approximately)	Distance on time line	Period
Kaibab Formation	Ended 255,000,000 years ago Began 260,000,000 years ago	25.5 cm 26.0 cm	
Toroweap Formation	Ended 260,000,000 years ago Began 265,000,000 years ago	26.0 cm 26.5 cm	
Coconino Sandstone	Ended 265,000,000 years ago Began 270,000,000 years ago	26.5 cm 27.0 cm	
Hermit Shale	Ended 270,000,000 years ago Began 275,000,000 years ago	27.0 cm 27.5 cm	
Supai Group	Ended 275,000,000 years ago Began 325,000,000 years ago	27.5 cm 32.5 cm	
Redwall Limestone	Ended 325,000,000 years ago Began 360,000,000 years ago	32.5 cm 36.0 cm	
Temple Butte Limestone	Ended 370,000,000 years ago Began 375,000,000 years ago	37.0 cm 37.5 cm	
Muav Limestone	Ended 525,000,000 years ago Began 530,000,000 years ago	52.5 cm 53.0 cm	
Bright Angel Shale	Ended 530,000,000 years ago Began 540,000,000 years ago	53.0 cm 54.0 cm	
Tapeats Sandstone	Ended 540,000,000 years ago Began 545,000,000 years ago	54.0 cm 54.5 cm	

1 cm = 10,000,000 years

FOSS Next Generation
© The Regents of the University of California
Can be duplicated for classroom or workshop use.
Earth History Course
Investigation 4: Fossils and Past Environments
No. 28—Notebook Master

Cooling-Rate Investigation

WARNING — This set contains chemicals that may be harmful if misused. Read cautions on individual containers carefully. Not to be used by children except under adult supervision.

What effect does cooling rate have on crystal formation?

1. What variables do you need to consider in the design of your experiment?

2. What materials will you need for your investigation?

3. Describe your procedure.

4. Record data and/or describe your results.

5. What conclusions can you draw about crystals in igneous rocks?

Cooling-Rate Investigation

WARNING — This set contains chemicals that may be harmful if misused. Read cautions on individual containers carefully. Not to be used by children except under adult supervision.

What effect does cooling rate have on crystal formation?

1. What variables do you need to consider in the design of your experiment?

2. What materials will you need for your investigation?

3. Describe your procedure.

4. Record data and/or describe your results.

5. What conclusions can you draw about crystals in igneous rocks?

FOSS Next Generation
© The Regents of the University of California
Can be duplicated for classroom or workshop use.
Earth History Course
Investigation 5: Igneous Rocks
No. 29—Notebook Master

Igneous-Rock Observations

WARNING — This set contains chemicals that may be harmful if misused. Read cautions on individual containers carefully. Not to be used by children except under adult supervision.

Rock	Description	Intrusive or extrusive?	Identification
11			
12			
16			
17			
19			
21			
23			
24			

Igneous-Rock Observations

WARNING — This set contains chemicals that may be harmful if misused. Read cautions on individual containers carefully. Not to be used by children except under adult supervision.

Rock	Description	Intrusive or extrusive?	Identification
11			
12			
16			
17			
19			
21			
23			
24			

FOSS Next Generation
© The Regents of the University of California
Can be duplicated for classroom or workshop use.
Earth History Course
Investigation 5: Igneous Rocks
No. 30—Notebook Master

Rock-Layer Age Puzzle

This illustration shows a rock column. Using potassium-argon dating, geologists have calculated an age of 200 million years for rock A, a granite. Rock F, the volcano, has been given an age of 225,000 years. Use the ages and illustration to answer the questions.

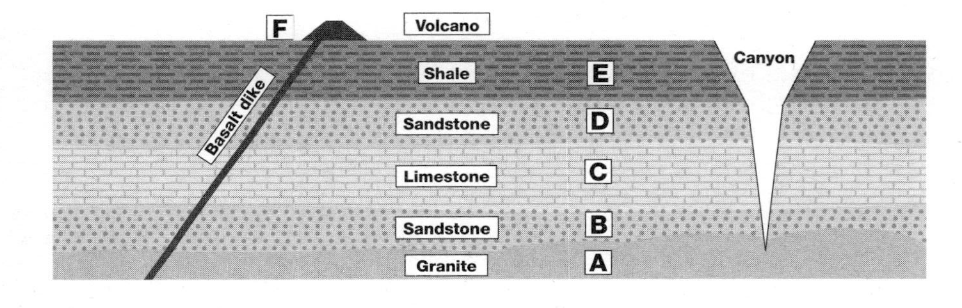

1. Which rock (A, B, C, D, E, or F) is the oldest? How do you know?

2. Which rock (A, B, C, D, E, or F) is the youngest? How do you know?

3. Did the canyon form before or after layers B, C, D, and E? How do you know?

4. Did rock B form before or after rock C? How do you know?

5. When did rocks B, C, D, and E form? Give a range, between XXX and XXX. How do you know?

6. Which rock layers are sedimentary rocks? Which are igneous rocks?

Rock-Layer Age Puzzle

This illustration shows a rock column. Using potassium-argon dating, geologists have calculated an age of 200 million years for rock A, a granite. Rock F, the volcano, has been given an age of 225,000 years. Use the ages and illustration to answer the questions.

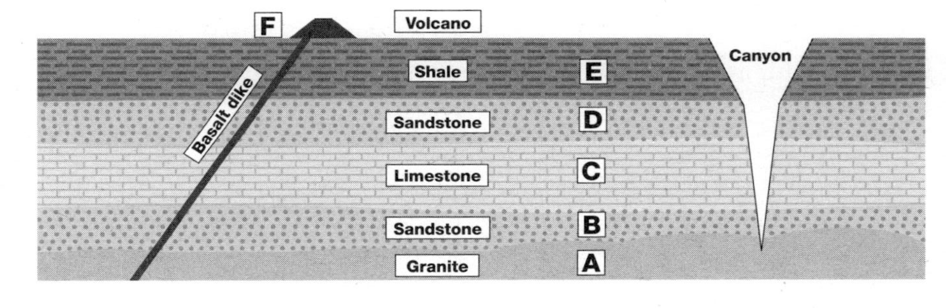

1. Which rock (A, B, C, D, E, or F) is the oldest? How do you know?

2. Which rock (A, B, C, D, E, or F) is the youngest? How do you know?

3. Did the canyon form before or after layers B, C, D, and E? How do you know?

4. Did rock B form before or after rock C? How do you know?

5. When did rocks B, C, D, and E form? Give a range, between XXX and XXX. How do you know?

6. Which rock layers are sedimentary rocks? Which are igneous rocks?

FOSS Next Generation
© The Regents of the University of California
Can be duplicated for classroom or workshop use.
Earth History Course
Investigation 5: Igneous Rocks
No. 31—Notebook Master

"Wegener" Questions

1. What evidence did Wegener use to support his idea of continental drift?

2. What did other scientists say about Wegener's ideas?

3. How did other scientists explain why the continents seemed to fit together?

"Wegener" Questions

1. What evidence did Wegener use to support his idea of continental drift?

2. What did other scientists say about Wegener's ideas?

3. How did other scientists explain why the continents seemed to fit together?

FOSS Next Generation
© The Regents of the University of California
Can be duplicated for classroom or workshop use.
Earth History Course
Investigation 6: Volcanoes and Earthquakes
No. 32—Notebook Master

Earth's Layers Information

Layer	Consistency	Composition	Temperature	Density	Thickness
		Basalt	From ocean temperature to 870°C near the mantle boundary	More dense than the continental crust	5 km to 8 km
		Igneous rocks like granite, sedimentary rocks, other rocks	From air temperature to 870°C near the mantle boundary	Less dense than the oceanic crust	30 km to 60 km
		More iron and magnesium than the crust; less silicon and aluminum	500°C to 900°C	More dense than crust and asthenosphere	0 km to 200 km
		More iron, magnesium, and calcium than the crust	500°C near crust to 4000°C near core	More dense than crust, less dense than solid upper mantle	2,900 km
		Iron, nickel	4400°C to 6100°C	More dense than the mantle	2,200 km
		Iron, nickel	7000°C	Most dense layer	1,250 km in diameter

Earth's Layers Information

Layer	Consistency	Composition	Temperature	Density	Thickness
		Basalt	From ocean temperature to 870°C near the mantle boundary	More dense than the continental crust	5 km to 8 km
		Igneous rocks like granite, sedimentary rocks, other rocks	From air temperature to 870°C near the mantle boundary	Less dense than the oceanic crust	30 km to 60 km
		More iron and magnesium than the crust; less silicon and aluminum	500°C to 900°C	More dense than crust and asthenosphere	0 km to 200 km
		More iron, magnesium, and calcium than the crust	500°C near crust to 4000°C near core	More dense than crust, less dense than solid upper mantle	2,900 km
		Iron, nickel	4400°C to 6100°C	More dense than the mantle	2,200 km
		Iron, nickel	7000°C	Most dense layer	1,250 km in diameter

FOSS Next Generation
© The Regents of the University of California
Can be duplicated for classroom or workshop use.

Earth History Course
Investigation 6: Volcanoes and Earthquakes
No. 33—Notebook Master

The History of the Theory of Plate Tectonics

Think about a time before seafloor maps and satellite photos of Earth, even before accurate global land maps.

The idea that Earth's outer layer is made of moving **plates** was not widely accepted until the 1960s. The theory of **plate tectonics** was probably not included in your grandparents' science textbooks. The theory is built on centuries of data and scientific development.

Geologic Puzzles

The origins of the theory go back to the first world maps in the late 1500s. These maps included most of Earth. After seeing the shapes of Africa and South America, some people wondered if the two continents were once connected. But how could that have happened? It took 300 years for scientists to come up with some ideas for how continents move.

As explorers traveled to the far reaches of Earth, they asked, How did fossils of sea creatures get on top of tall mountains? Is there a relationship between volcanoes and earthquakes? Were the continents once close together, making one big landform?

We now know that forces inside Earth are continually reshaping Earth's surface. As an oceanic volcano spews ash and cinders and sends rivers of lava to the ocean, it is creating new land.

The History of the Theory of Plate Tectonics

Think about a time before seafloor maps and satellite photos of Earth, even before accurate global land maps.

The idea that Earth's outer layer is made of moving **plates** was not widely accepted until the 1960s. The theory of **plate tectonics** was probably not included in your grandparents' science textbooks. The theory is built on centuries of data and scientific development.

Geologic Puzzles

The origins of the theory go back to the first world maps in the late 1500s. These maps included most of Earth. After seeing the shapes of Africa and South America, some people wondered if the two continents were once connected. But how could that have happened? It took 300 years for scientists to come up with some ideas for how continents move.

As explorers traveled to the far reaches of Earth, they asked, How did fossils of sea creatures get on top of tall mountains? Is there a relationship between volcanoes and earthquakes? Were the continents once close together, making one big landform?

We now know that forces inside Earth are continually reshaping Earth's surface. As an oceanic volcano spews ash and cinders and sends rivers of lava to the ocean, it is creating new land.

74

FOSS Next Generation
© The Regents of the University of California
Can be duplicated for classroom or workshop use.

Earth History Course
Investigation 6: Volcanoes and Earthquakes
No. 34—Notebook Master

Explaining the Puzzle

There were two main explanations for the mountaintop marine fossils. Some believed that global flooding raised the sea level above the highest peaks of the world. But otherwise, Earth's land had never changed. Others observed earthquakes and volcanic activity. They reasoned that processes inside Earth changed the surface, creating new hills and mountains.

James Hutton (1726–1797), a Scottish geologist, supported this second explanation. He observed that streams carried sediments away from his farm. So why had erosion not made the world into a perfectly round sphere? He decided that forces lifting sections of Earth's surface must balance out erosion. Hutton's theory required large amounts of heat energy from inside Earth and extremely long periods of time.

These were brilliant new ideas about Earth's history. But Hutton was a poor writer. Even the brightest scientific minds could not understand his written explanations.

After Hutton died, a close friend rewrote his book about geology. Eventually, scientists accepted Hutton's ideas of a changing planet. But understanding the evidence for plate tectonics was a long way off.

World maps in the 1550s began to give a more complete view of Earth. Some people began to wonder if South America and Africa had once been connected.

Explaining the Puzzle

There were two main explanations for the mountaintop marine fossils. Some believed that global flooding raised the sea level above the highest peaks of the world. But otherwise, Earth's land had never changed. Others observed earthquakes and volcanic activity. They reasoned that processes inside Earth changed the surface, creating new hills and mountains.

James Hutton (1726–1797), a Scottish geologist, supported this second explanation. He observed that streams carried sediments away from his farm. So why had erosion not made the world into a perfectly round sphere? He decided that forces lifting sections of Earth's surface must balance out erosion. Hutton's theory required large amounts of heat energy from inside Earth and extremely long periods of time.

These were brilliant new ideas about Earth's history. But Hutton was a poor writer. Even the brightest scientific minds could not understand his written explanations.

After Hutton died, a close friend rewrote his book about geology. Eventually, scientists accepted Hutton's ideas of a changing planet. But understanding the evidence for plate tectonics was a long way off.

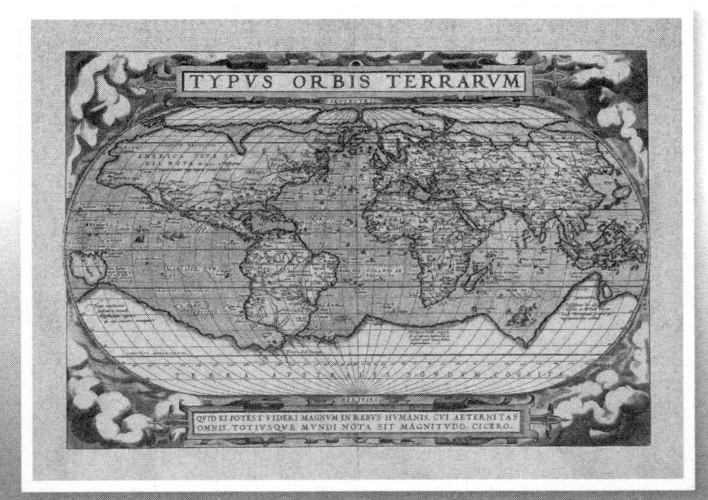

World maps in the 1550s began to give a more complete view of Earth. Some people began to wonder if South America and Africa had once been connected.

Investigation 6: Volcanoes and Earthquakes 75

FOSS Next Generation
© The Regents of the University of California
Can be duplicated for classroom or workshop use.

Earth History Course
Investigation 6: Volcanoes and Earthquakes
No. 35—Notebook Master

Putting the Pieces Together

The puzzle-like shapes of the continents intrigued Alfred Wegener (1880–1930), a German meteorologist. He was also interested in odd connections among fossils. For example, he found fossils of animals that once lived in tropical climates in areas that now have cold climates. He observed that the same plants and animals appeared as fossils in rocks of the same period on different sides of the ocean. The fossils included a freshwater reptile that was like a small crocodile found in Brazil and South Africa.

Can you imagine an entire community of reptiles traveling from Africa to South America? Neither could Wegener. Instead, he proposed a world where all the continents were connected as one huge continent. He called it Pangaea. He wrote of Pangaea as a land "where flora and fauna were able to mingle together before they were split apart." In the early 1900s, he published his idea of drifting continents, a new way of viewing Earth's history. But at the time, most scientists believed the continents were anchored in place. The continents might move up and down, but they certainly did not drift around the planet.

Earth's Landmasses in Ancient Position

The color splashes show the possible patterns of fossil distribution before the continents split apart.

Putting the Pieces Together

The puzzle-like shapes of the continents intrigued Alfred Wegener (1880–1930), a German meteorologist. He was also interested in odd connections among fossils. For example, he found fossils of animals that once lived in tropical climates in areas that now have cold climates. He observed that the same plants and animals appeared as fossils in rocks of the same period on different sides of the ocean. The fossils included a freshwater reptile that was like a small crocodile found in Brazil and South Africa.

Can you imagine an entire community of reptiles traveling from Africa to South America? Neither could Wegener. Instead, he proposed a world where all the continents were connected as one huge continent. He called it Pangaea. He wrote of Pangaea as a land "where flora and fauna were able to mingle together before they were split apart." In the early 1900s, he published his idea of drifting continents, a new way of viewing Earth's history. But at the time, most scientists believed the continents were anchored in place. The continents might move up and down, but they certainly did not drift around the planet.

Earth's Landmasses in Ancient Position

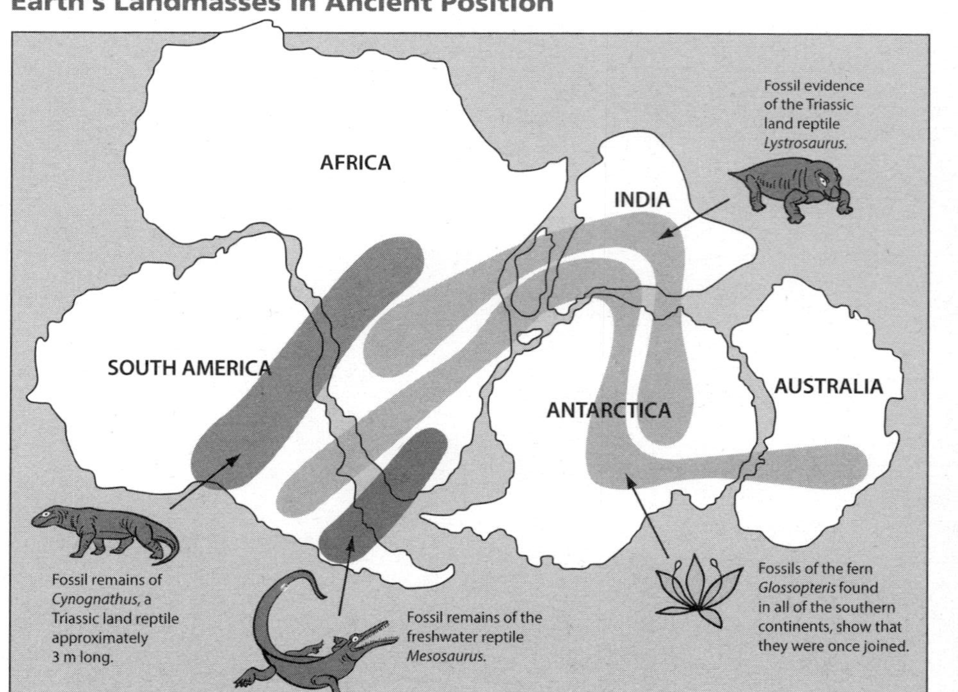

The color splashes show the possible patterns of fossil distribution before the continents split apart.

76

FOSS Next Generation
© The Regents of the University of California
Can be duplicated for classroom or workshop use.

Earth History Course
Investigation 6: Volcanoes and Earthquakes
No. 36—Notebook Master

Earth Maps over Time

PERMIAN
250 million years ago

TRIASSIC
200 million years ago

JURASSIC
145 million years ago

CRETACEOUS
65 million years ago

PRESENT DAY

This series of maps shows the probable positions of the continents following the breakup of the supercontinent Pangaea more than 200 million years ago.

Resistance to Continental Drift

Scientists dismissed Wegener's ideas as physically impossible. In fact, Wegener could not explain what forces had moved large masses of solid rock over such great distances. How could it have happened? Wegener fought for his ideas of moving continents and a more dynamic planet until his death in 1930.

Scientists found it hard to ignore the fossil evidence for Wegener's theory. To get around this problem, geologists imagined land bridges crossing the ocean. Fossils of an ancient horse were found in France and Florida. So a land bridge was drawn across the Atlantic Ocean. A lost continent was an attempt to explain connections that spanned the Indian Ocean. These land bridges disappeared when technology allowed us to map the seafloor.

The Missing Pieces

In the 16th century, sailors used long ropes to measure the depth of the ocean. They found out that the seafloor is not as flat as people thought. Knowledge of the topography of the seafloor increased with the development of modern tools. In 1853, US Navy lieutenant Matthew Maury (1806–1873) published the first ocean-bottom chart. It showed evidence of underwater mountains in the central Atlantic Ocean. Survey ships laying the trans-Atlantic telegraph cable confirmed Maury's findings.

Earth Maps over Time

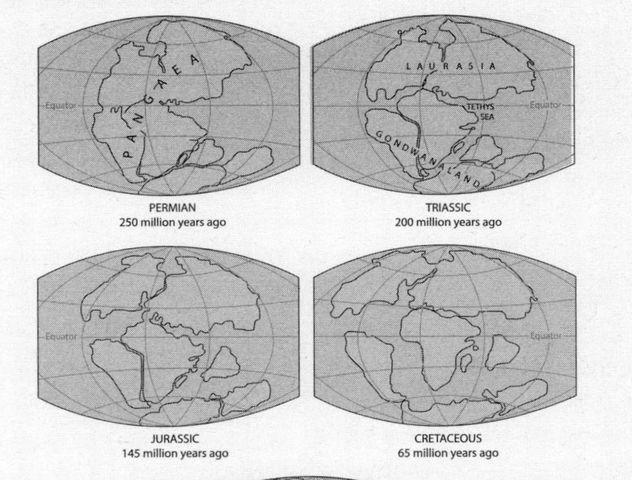

PERMIAN
250 million years ago

TRIASSIC
200 million years ago

JURASSIC
145 million years ago

CRETACEOUS
65 million years ago

PRESENT DAY

This series of maps shows the probable positions of the continents following the breakup of the supercontinent Pangaea more than 200 million years ago.

Resistance to Continental Drift

Scientists dismissed Wegener's ideas as physically impossible. In fact, Wegener could not explain what forces had moved large masses of solid rock over such great distances. How could it have happened? Wegener fought for his ideas of moving continents and a more dynamic planet until his death in 1930.

Scientists found it hard to ignore the fossil evidence for Wegener's theory. To get around this problem, geologists imagined land bridges crossing the ocean. Fossils of an ancient horse were found in France and Florida. So a land bridge was drawn across the Atlantic Ocean. A lost continent was an attempt to explain connections that spanned the Indian Ocean. These land bridges disappeared when technology allowed us to map the seafloor.

The Missing Pieces

In the 16th century, sailors used long ropes to measure the depth of the ocean. They found out that the seafloor is not as flat as people thought. Knowledge of the topography of the seafloor increased with the development of modern tools. In 1853, US Navy lieutenant Matthew Maury (1806–1873) published the first ocean-bottom chart. It showed evidence of underwater mountains in the central Atlantic Ocean. Survey ships laying the trans-Atlantic telegraph cable confirmed Maury's findings.

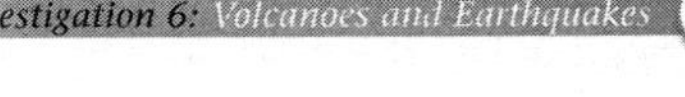

Investigation 6: Volcanoes and Earthquakes 77

FOSS Next Generation
© The Regents of the University of California
Can be duplicated for classroom or workshop use.

Earth History Course
Investigation 6: Volcanoes and Earthquakes
No. 37—Notebook Master

The first sonar systems were developed during World War I (1914–1918) to help locate German submarines. These systems recorded the time it took for sound to travel from the ship to the seafloor and back again. The ocean depth was calculated from these times. The data collected from this early sonar confirmed the existence of the Mid-Atlantic Ridge.

Sonar can also measure the thickness of sediments, an important tool in the study of the seafloor. In 1947, **seismologists** on the US research ship *Atlantis* measured the sediment layer in the Atlantic. Scientists had believed that the ocean was at least 4 billion years old. If it was that old, the sediment layer on the ocean bottom should have become very thick. But evidence from sonar readings showed that the layer was relatively thin. Why was that?

Seafloor Spreading

Harry Hammond Hess (1906–1969) was the geologist who finally came up with evidence that supported Wegener's theory of drifting continents. During World War II (1939–1945), Hess was captain of a transport ship equipped with the newest form of sonar. He decided to use the equipment to gather data all the time. He noticed that the sediment layer was thinnest near the mid-ocean ridge. It got thicker as he traveled away from the ridge. The sediments nearest the ridge had been deposited over less time than those farther from it. From this evidence, Hess and other scientists inferred that the crust nearer to the ridge was younger. This difference in age suggests that the Atlantic seafloor is spreading. It is pushing Africa and South America apart by about 5 centimeters (cm) a year.

Mid-Atlantic Ridge

The Mid-Atlantic Ridge is part of the underwater mountain range that winds around the globe for about 70,000 km.

The first sonar systems were developed during World War I (1914–1918) to help locate German submarines. These systems recorded the time it took for sound to travel from the ship to the seafloor and back again. The ocean depth was calculated from these times. The data collected from this early sonar confirmed the existence of the Mid-Atlantic Ridge.

Sonar can also measure the thickness of sediments, an important tool in the study of the seafloor. In 1947, **seismologists** on the US research ship *Atlantis* measured the sediment layer in the Atlantic. Scientists had believed that the ocean was at least 4 billion years old. If it was that old, the sediment layer on the ocean bottom should have become very thick. But evidence from sonar readings showed that the layer was relatively thin. Why was that?

Seafloor Spreading

Harry Hammond Hess (1906–1969) was the geologist who finally came up with evidence that supported Wegener's theory of drifting continents. During World War II (1939–1945), Hess was captain of a transport ship equipped with the newest form of sonar. He decided to use the equipment to gather data all the time. He noticed that the sediment layer was thinnest near the mid-ocean ridge. It got thicker as he traveled away from the ridge. The sediments nearest the ridge had been deposited over less time than those farther from it. From this evidence, Hess and other scientists inferred that the crust nearer to the ridge was younger. This difference in age suggests that the Atlantic seafloor is spreading. It is pushing Africa and South America apart by about 5 centimeters (cm) a year.

Mid-Atlantic Ridge

The Mid-Atlantic Ridge is part of the underwater mountain range that winds around the globe for about 70,000 km.

78

FOSS Next Generation
© The Regents of the University of California
Can be duplicated for classroom or workshop use.
Earth History Course
Investigation 6: Volcanoes and Earthquakes
No. 38—Notebook Master

Today, we combine satellite images of underwater mountains with earthquake and volcano data to study the processes that shape Earth. These data are evidence that Earth's crust is composed of solid **tectonic plates**. They float on the fluid portion of the **mantle**. This evidence strongly supports the modern theory of moving crustal plates, called plate tectonics. It took courage and conviction for scientists like Wegener and Hess to defend and promote their unconventional ideas. Their new ideas changed the way everyone thinks about Earth.

Think Questions

1. **What evidence caused Wegener to think the continents had been connected at one time?**
2. **Why did most geologists disagree with Wegener's ideas?**
3. **What are two pieces of evidence that scientists used to confirm the theory of plate tectonics?**

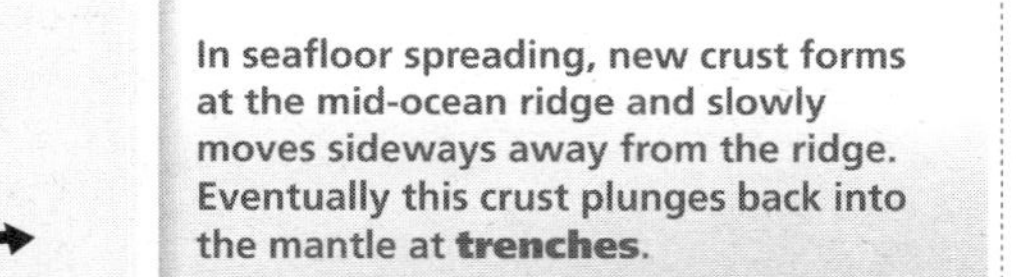

In seafloor spreading, new crust forms at the mid-ocean ridge and slowly moves sideways away from the ridge. Eventually this crust plunges back into the mantle at **trenches**.

The Mid-Atlantic Ridge—the boundary between the North American and Eurasian tectonic plates—slices through the center of Iceland. Rocky outcroppings mark where the plates are slowly pulling apart.

Today, we combine satellite images of underwater mountains with earthquake and volcano data to study the processes that shape Earth. These data are evidence that Earth's crust is composed of solid **tectonic plates**. They float on the fluid portion of the **mantle**. This evidence strongly supports the modern theory of moving crustal plates, called plate tectonics. It took courage and conviction for scientists like Wegener and Hess to defend and promote their unconventional ideas. Their new ideas changed the way everyone thinks about Earth.

Think Questions

1. **What evidence caused Wegener to think the continents had been connected at one time?**
2. **Why did most geologists disagree with Wegener's ideas?**
3. **What are two pieces of evidence that scientists used to confirm the theory of plate tectonics?**

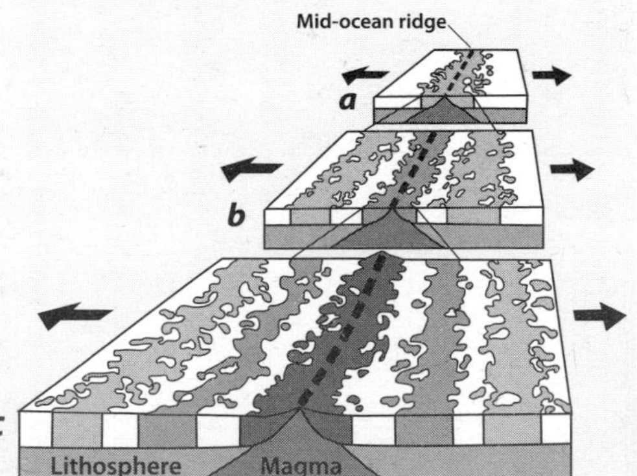

In seafloor spreading, new crust forms at the mid-ocean ridge and slowly moves sideways away from the ridge. Eventually this crust plunges back into the mantle at **trenches**.

The Mid-Atlantic Ridge—the boundary between the North American and Eurasian tectonic plates—slices through the center of Iceland. Rocky outcroppings mark where the plates are slowly pulling apart.

Investigation 6: Volcanoes and Earthquakes 79

FOSS Next Generation
© The Regents of the University of California
Can be duplicated for classroom or workshop use.

Earth History Course
Investigation 6: Volcanoes and Earthquakes
No. 39—Notebook Master

Plate Boundaries

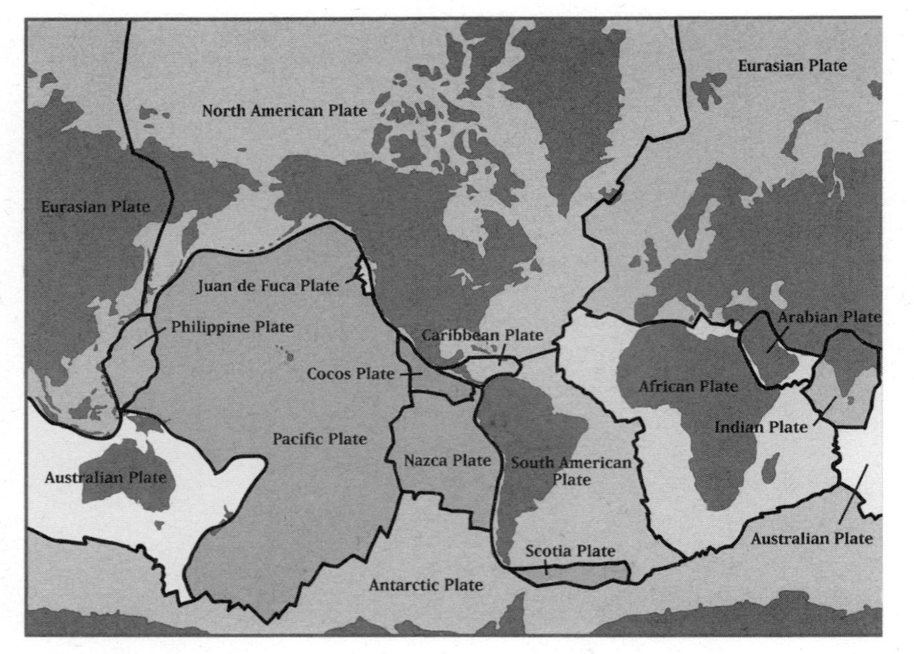

1. Open "Plate Boundaries" in Google Earth™.
2. In your notebook, make a chart with three columns: "Name," "Type of Boundary," and "Landforms."
3. Pick five sites from the possible list (you can click each site to visit it) and record them in the "Name" column.
4. Using the information in Google Earth™, record the type of plate boundary (divergent, transform, or convergent) in the "Type of Boundary" column.
5. For each location, mark and label the map above for that site, and draw arrows to show how the plates are moving relative to each other.
6. Record your observations about the landforms and other features you observe at that site in the "Landforms" column.

Plate Boundaries

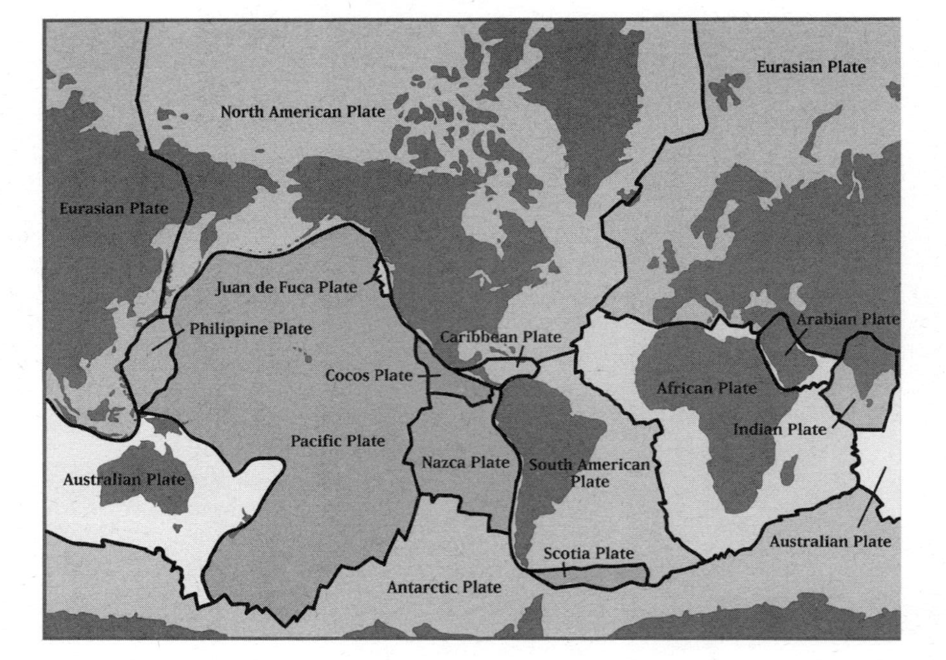

1. Open "Plate Boundaries" in Google Earth™.
2. In your notebook, make a chart with three columns: "Name," "Type of Boundary," and "Landforms."
3. Pick five sites from the possible list (you can click each site to visit it) and record them in the "Name" column.
4. Using the information in Google Earth™, record the type of plate boundary (divergent, transform, or convergent) in the "Type of Boundary" column.
5. For each location, mark and label the map above for that site, and draw arrows to show how the plates are moving relative to each other.
6. Record your observations about the landforms and other features you observe at that site in the "Landforms" column.

FOSS Next Generation
© The Regents of the University of California
Can be duplicated for classroom or workshop use.
Earth History Course
Investigation 6: Volcanoes and Earthquakes
No. 40—Notebook Master

Plate-Interaction Map

Based on http://denali.gsfc.nasa.gov/dtam

Plate-Interaction Map

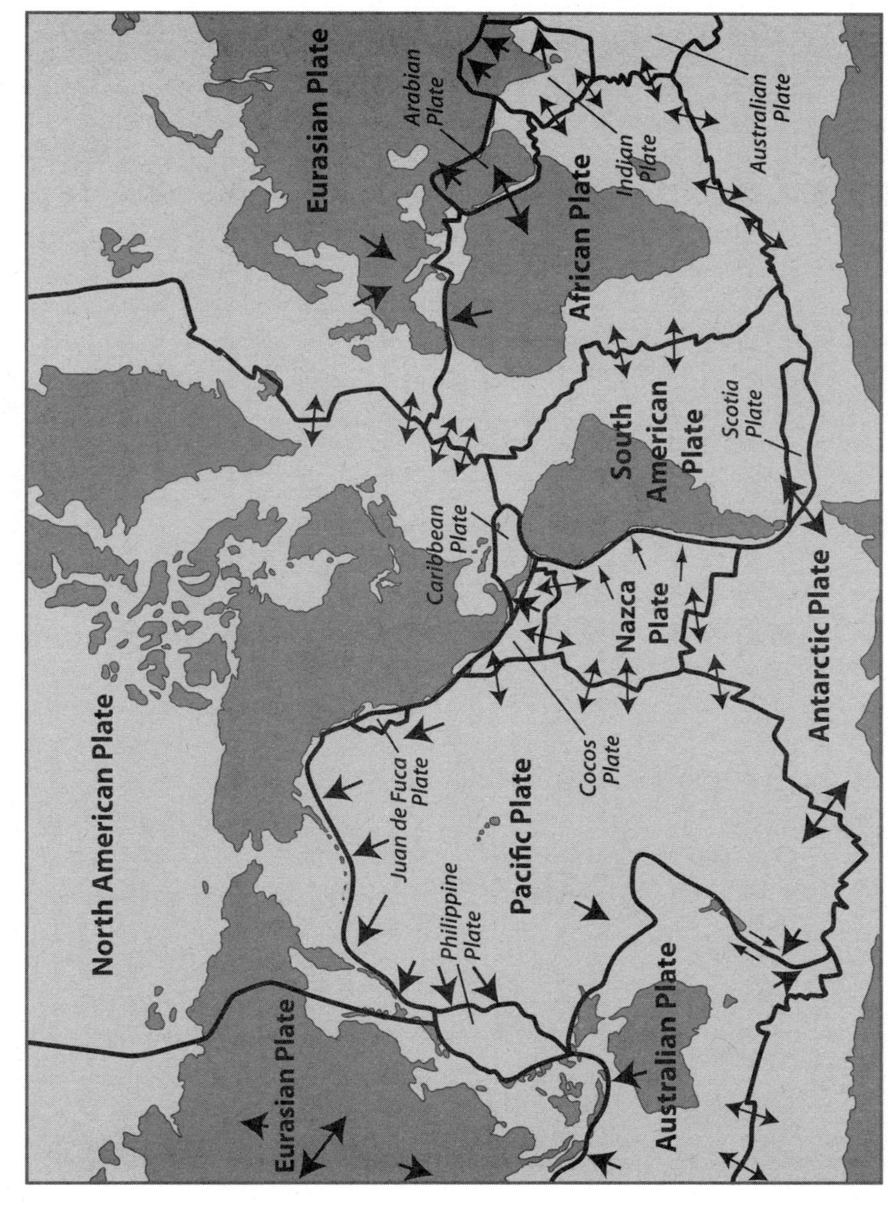

Based on http://denali.gsfc.nasa.gov/dtam

FOSS Next Generation
© The Regents of the University of California
Can be duplicated for classroom or workshop use.

Earth History Course
Investigation 6: Volcanoes and Earthquakes
No. 41—Notebook Master

Mountain Types

Type	How it forms	Drawing	Example
Fold	Slow compression caused by converging plates.		
Fault block	Breaking of the crust either by compression or pulling apart (tension). The movement happens fast enough to break, rather than fold, the crust.		
Dome	Magma pushing up from below uplifts the rocks. Erosion can expose the hardened magma dome.		
Volcano	Volcanic eruption.		
Plateau	Uplift of a relatively flat area next to a mountain that is being created by folding.		

Mountain Types

Type	How it forms	Drawing	Example
Fold	Slow compression caused by converging plates.		
Fault block	Breaking of the crust either by compression or pulling apart (tension). The movement happens fast enough to break, rather than fold, the crust.		
Dome	Magma pushing up from below uplifts the rocks. Erosion can expose the hardened magma dome.		
Volcano	Volcanic eruption.		
Plateau	Uplift of a relatively flat area next to a mountain that is being created by folding.		

FOSS Next Generation
© The Regents of the University of California
Can be duplicated for classroom or workshop use.
Earth History Course
Investigation 7: Mountains and Metamorphic Rocks
No. 42—Notebook Master

Shenandoah Field Trip

Mile 5.7
Signal Knob Overlook

Mile 21
Hogback Overlook

Mile 33.1
Hazel Mountain Overlook

Mile 35.1
Pinnacles Overlook

Mile 38.6
Stony Man Overlook

Mile 39.1
Little Stony Man Trail

Mile 49
Franklin Cliffs Overlook

Mile 51.2
Big Meadows area, Blackrock

Shenandoah Field Trip

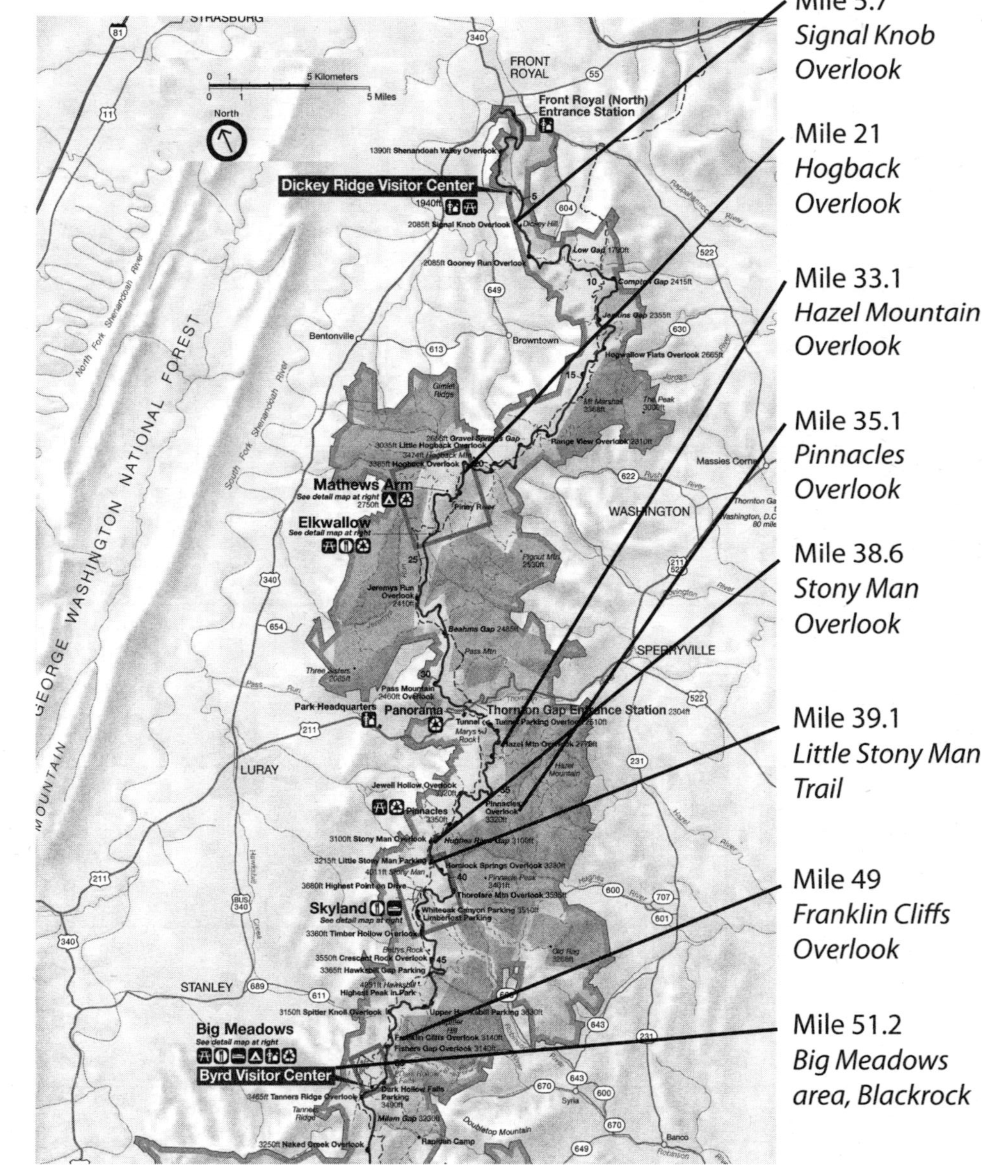

Mile 5.7
Signal Knob Overlook

Mile 21
Hogback Overlook

Mile 33.1
Hazel Mountain Overlook

Mile 35.1
Pinnacles Overlook

Mile 38.6
Stony Man Overlook

Mile 39.1
Little Stony Man Trail

Mile 49
Franklin Cliffs Overlook

Mile 51.2
Big Meadows area, Blackrock

FOSS Next Generation
© The Regents of the University of California
Can be duplicated for classroom or workshop use.

Earth History Course
Investigation 7: Mountains and Metamorphic Rocks
No. 43—Notebook Master

Field-Trip Notes

Mile 5.7 *Signal Knob Overlook*	
Mile 21 *Hogback Overlook*	
Mile 33.1 *Hazel Mountain Overlook*	
Mile 35.1 *Pinnacles Overlook*	
Mile 38.6 *Stony Man Overlook*	
Mile 39.1 *Little Stony Man Trail*	
Mile 49 *Franklin Cliffs Overlook*	
Mile 51.2 *Big Meadows area, Blackrock*	

Field-Trip Notes

Mile 5.7 *Signal Knob Overlook*	
Mile 21 *Hogback Overlook*	
Mile 33.1 *Hazel Mountain Overlook*	
Mile 35.1 *Pinnacles Overlook*	
Mile 38.6 *Stony Man Overlook*	
Mile 39.1 *Little Stony Man Trail*	
Mile 49 *Franklin Cliffs Overlook*	
Mile 51.2 *Big Meadows area, Blackrock*	

FOSS Next Generation
© The Regents of the University of California
Can be duplicated for classroom or workshop use.
Earth History Course
Investigation 7: Mountains and Metamorphic Rocks
No. 44—Notebook Master

Geoscenario Team Questions

Assignment

Use the following questions to collect information to tell the story of a geoscenario. Then work together as a team to combine notes and create one presentation that you will each be able to share with your classmates in small groups.

Use a page in your notebook to take notes on each of the following questions and a page or two for notes on your final team presentation.

Project guiding questions

1. What geological features define your geoscenario?
2. What earth processes formed your geoscenario?
3. How is your geoscenario significant to humans' future?
4. What important events are in the time line of the geological and human stories of this geoscenario?

Presentation guidelines

When using articles, charts, graphs, and pictures in your presentation, remember these guidelines.

1. Convert any written research material into drawings, charts, tables, or bullets and/or completely rewrite the text in your own words.
2. Cite the source of the information you use.
3. Create an informative caption for all visuals (pictures, charts, or graphs), written in complete sentences and your own words, and cite the source of the image.

Specialist role:

Geoscenario Team Questions

Assignment

Use the following questions to collect information to tell the story of a geoscenario. Then work together as a team to combine notes and create one presentation that you will each be able to share with your classmates in small groups.

Use a page in your notebook to take notes on each of the following questions and a page or two for notes on your final team presentation.

Project guiding questions

1. What geological features define your geoscenario?
2. What earth processes formed your geoscenario?
3. How is your geoscenario significant to humans' future?
4. What important events are in the time line of the geological and human stories of this geoscenario?

Presentation guidelines

When using articles, charts, graphs, and pictures in your presentation, remember these guidelines.

1. Convert any written research material into drawings, charts, tables, or bullets and/or completely rewrite the text in your own words.
2. Cite the source of the information you use.
3. Create an informative caption for all visuals (pictures, charts, or graphs), written in complete sentences and your own words, and cite the source of the image.

Specialist role:

FOSS Next Generation
© The Regents of the University of California
Can be duplicated for classroom or workshop use.
Earth History Course
Investigation 8: Geoscenarios
No. 45—Notebook Master

Geoscenario Work Checklist

Geoscenario name: ________________________________

Preparing information

Review this checklist as you gather information and prepare for your presentation. Initial each step as you complete it.

____ I collected notes for the four guiding questions from the introduction and specialist articles.

____ I worked with my team to combine our notes.

____ I worked with my team to write notes for a presentation of our information in an engaging, effective manner.

____ I used visuals, and I referenced evidence to support statements.

____ All my writing in the final project is in my own words.

____ I understand my teammates' parts well enough to answer questions other students may ask about any part of our geoscenario.

____ We have decided who is presenting which information.

____ We have rehearsed our presentation.

Presenting information

Use your visuals and evidence during your presentation to illustrate your ideas, and after your presentation to help answer questions.

Gaining information from other groups

When you are learning about other geoscenarios, ask questions and record information to address the four guiding questions. Support your answers with evidence presented by the other teams.

Reflecting on the project

On the next notebook page, write a paragraph on why you think knowing the geological story of a place is important to humans.

Geoscenario Work Checklist

Geoscenario name: ________________________________

Preparing information

Review this checklist as you gather information and prepare for your presentation. Initial each step as you complete it.

____ I collected notes for the four guiding questions from the introduction and specialist articles.

____ I worked with my team to combine our notes.

____ I worked with my team to write notes for a presentation of our information in an engaging, effective manner.

____ I used visuals, and I referenced evidence to support statements.

____ All my writing in the final project is in my own words.

____ I understand my teammates' parts well enough to answer questions other students may ask about any part of our geoscenario.

____ We have decided who is presenting which information.

____ We have rehearsed our presentation.

Presenting information

Use your visuals and evidence during your presentation to illustrate your ideas, and after your presentation to help answer questions.

Gaining information from other groups

When you are learning about other geoscenarios, ask questions and record information to address the four guiding questions. Support your answers with evidence presented by the other teams.

Reflecting on the project

On the next notebook page, write a paragraph on why you think knowing the geological story of a place is important to humans.

FOSS Next Generation
© The Regents of the University of California
Can be duplicated for classroom or workshop use.

Earth History Course
Investigation 8: Geoscenarios
No. 46—Notebook Master

Reviewing Other Presentations

Follow these instructions for each geoscenario you visit.

1. On the *Geoscenario Presentation Notes* notebook sheet,
   - write the name of the geoscenario;
   - write the names of the experts presenting at the station.

2. Choose two of the following questions to ask the experts. Record the questions on *Geoscenario Presentation Notes*.
   a. What are interesting or important geologic features in this region?
   b. What geological processes formed this region?
   c. What can you say about the future of this region?
   d. How does the geology of this place affect people?
   e. How does the geology of this place affect the environment?

3. Create your own question for each geoscenario. Write the question on *Geoscenario Presentation Notes*.

4. Listen to the team's presentation. Ask your questions and record answers on *Geoscenario Presentation Notes*. Record additional notes on a labeled page in your notebook.

5. After each team's presentation, answer the last question on *Geoscenario Presentation Notes*: How is this location similar to or different from your group's site?

Reviewing Other Presentations

Follow these instructions for each geoscenario you visit.

1. On the *Geoscenario Presentation Notes* notebook sheet,
   - write the name of the geoscenario;
   - write the names of the experts presenting at the station.

2. Choose two of the following questions to ask the experts. Record the questions on *Geoscenario Presentation Notes*.
   a. What are interesting or important geologic features in this region?
   b. What geological processes formed this region?
   c. What can you say about the future of this region?
   d. How does the geology of this place affect people?
   e. How does the geology of this place affect the environment?

3. Create your own question for each geoscenario. Write the question on *Geoscenario Presentation Notes*.

4. Listen to the team's presentation. Ask your questions and record answers on *Geoscenario Presentation Notes*. Record additional notes on a labeled page in your notebook.

5. After each team's presentation, answer the last question on *Geoscenario Presentation Notes*: How is this location similar to or different from your group's site?

FOSS Next Generation
© The Regents of the University of California
Can be duplicated for classroom or workshop use.
Earth History Course
Investigation 8: Geoscenarios
No. 47—Notebook Master

Name ____________

Geoscenario Presentation Notes

Geoscenario: ____________

Presenting experts: ____________

Ask and answer questions from the *Reviewing Other Presentations* notebook sheet. Record the question letters in the spaces below.

Question ______

Question ______

My question (write it below)

Answer the following question after each team's presentation:

How is this location similar to or different from your group's site?

Name ____________

Geoscenario Presentation Notes

Geoscenario: ____________

Presenting experts: ____________

Ask and answer questions from the Reviewing Other Presentations notebook sheet. Record the question letters in the spaces below.

Question ______

Question ______

My question (write it below)

Answer the following question after each team's presentation:

How is this location similar to or different from your group's site?

FOSS Next Generation
© The Regents of the University of California
Can be duplicated for classroom or workshop use.
Earth History Course
Investigation 8: Geoscenarios
No. 48—Notebook Master

Grand Canyon Revisited A

Site 1

1. What type of rock is this? Cite evidence to support your statement.

2. What layer could it be? Cite evidence to support your statement.

3. How did this layer form?

4. How did this layer with sea fossils get to an elevation of 2,116 meters?

Site 2

1. What type of rock is this? Cite evidence to support your statement.

2. What layer could it be? Cite evidence to support your statement.

3. How did this layer form?

Site 3

1. What type of rock is this? Cite evidence to support your statement.

2. What layer could it be? Cite evidence to support your statement.

Grand Canyon Revisited A

Site 1

1. What type of rock is this? Cite evidence to support your statement.

2. What layer could it be? Cite evidence to support your statement.

3. How did this layer form?

4. How did this layer with sea fossils get to an elevation of 2,116 meters?

Site 2

1. What type of rock is this? Cite evidence to support your statement.

2. What layer could it be? Cite evidence to support your statement.

3. How did this layer form?

Site 3

1. What type of rock is this? Cite evidence to support your statement.

2. What layer could it be? Cite evidence to support your statement.

FOSS Next Generation
© The Regents of the University of California
Can be duplicated for classroom or workshop use.
Earth History Course
Investigation 9: What Is Earth's Story?
No. 49—Notebook Master

Grand Canyon Revisited B

Site 3 (continued)

3. How did this layer form?

Site 4

1. What type of rock is this? Cite evidence to support your statement.

2. How does this type of rock form?

3. How might this rock have ended up at the Grand Canyon?

Site 5

1. What type of rock is this? Cite evidence to support your statement.

2. How does this type of rock form?

3. How might this rock have ended up at the Grand Canyon?

Grand Canyon Revisited B

Site 3 (continued)

3. How did this layer form?

Site 4

1. What type of rock is this? Cite evidence to support your statement.

2. How does this type of rock form?

3. How might this rock have ended up at the Grand Canyon?

Site 5

1. What type of rock is this? Cite evidence to support your statement.

2. How does this type of rock form?

3. How might this rock have ended up at the Grand Canyon?

FOSS Next Generation
© The Regents of the University of California
Can be duplicated for classroom or workshop use.
Earth History Course
Investigation 9: What Is Earth's Story?
No. 50—Notebook Master

Teacher Masters

LABELS FOR ROCK BOXES

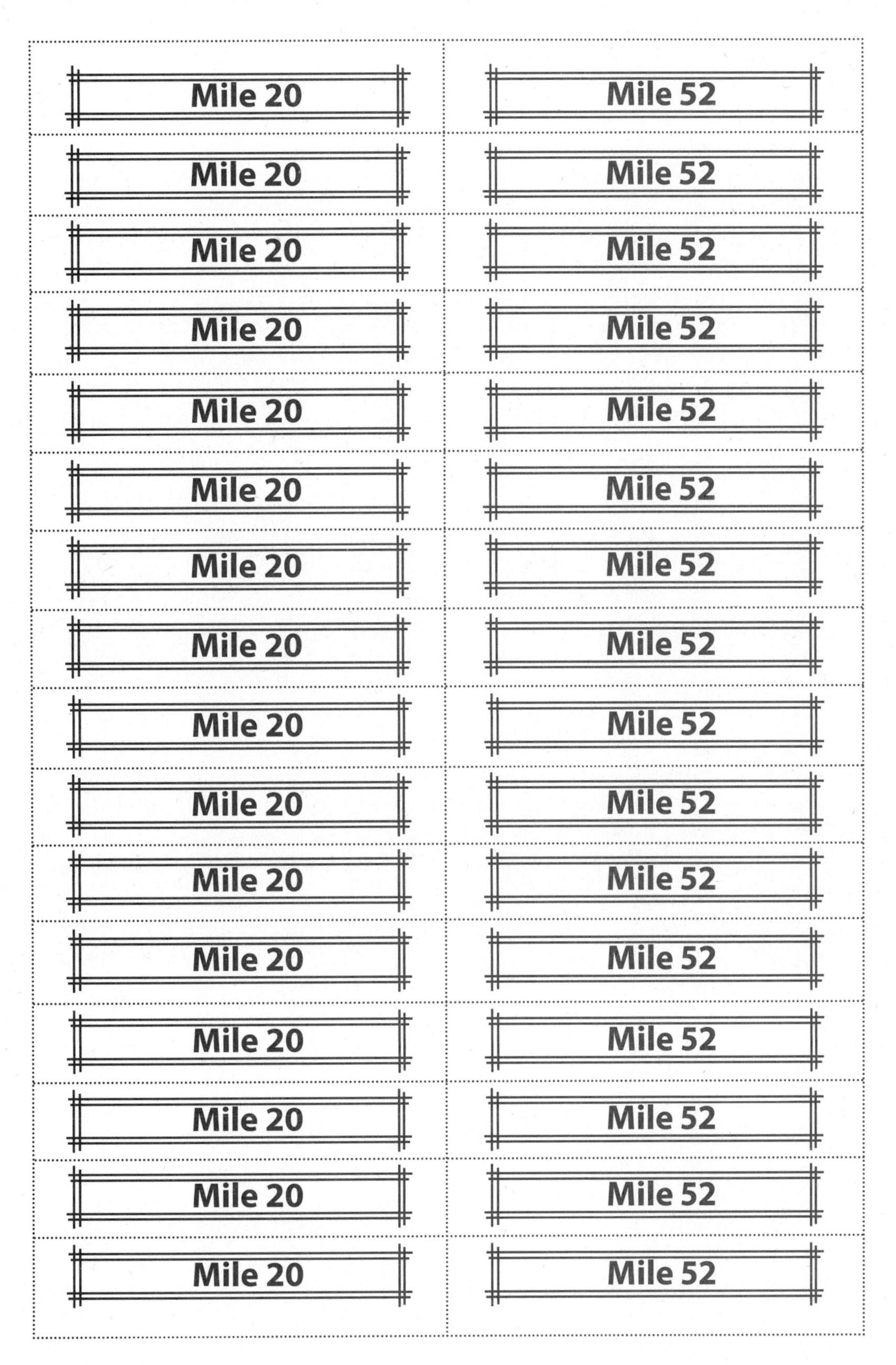

EIFFEL TOWER IN THE GRAND CANYON

GRAND CANYON MAP

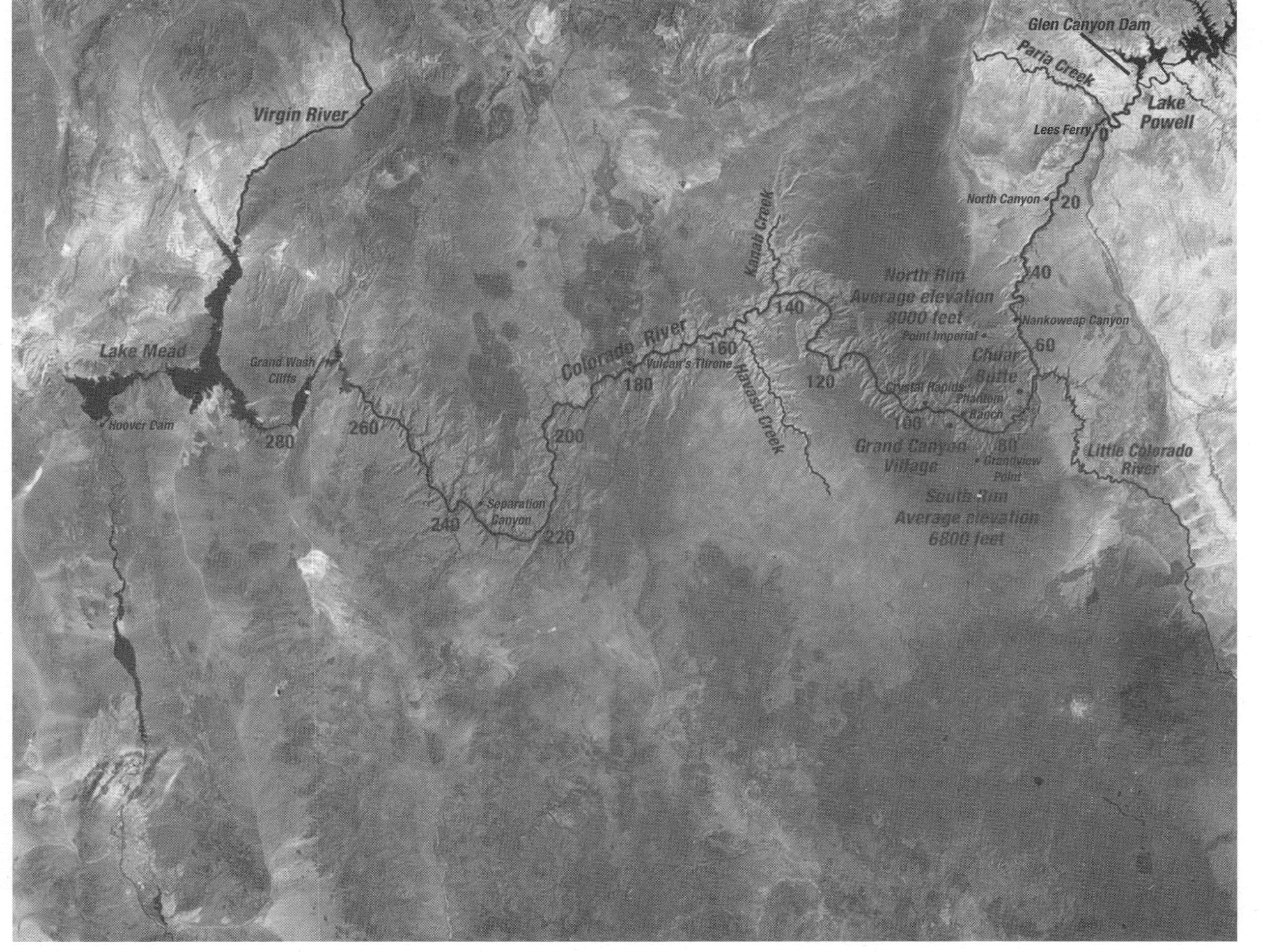

POWELL'S TECHNIQUE

ROCK LAYER KEY

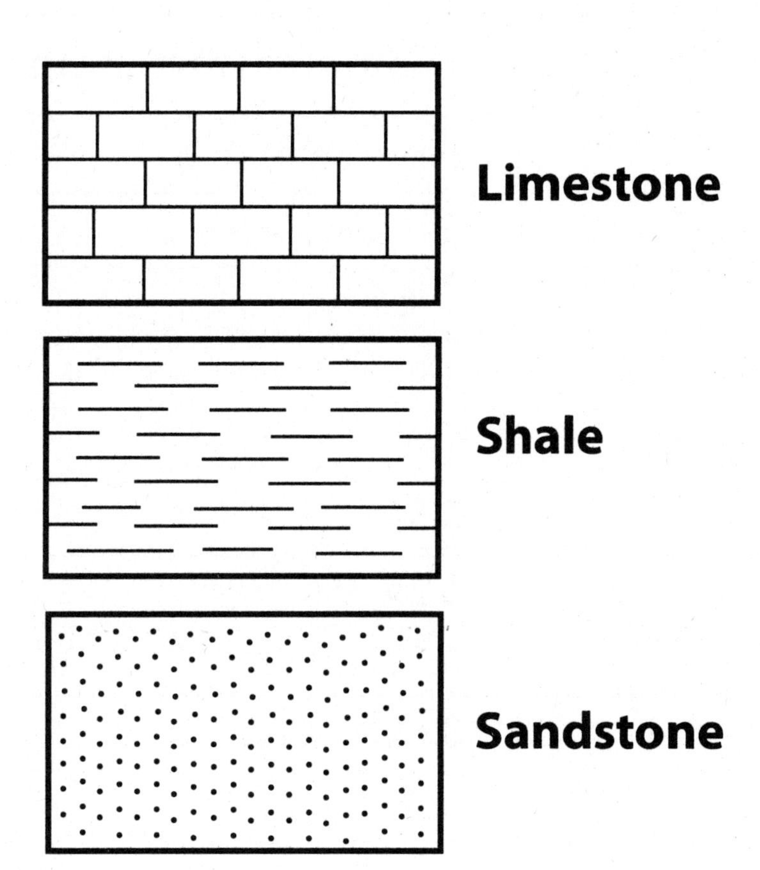

WENTWORTH SCALE OF ROCK PARTICLE SIZES

Classification	Diameter (millimeters)
Boulder	More than 256
Cobble	64–256
Pebble	4–64
Gravel (or granule)	2–4
Very coarse sand	1–2
Coarse sand	0.5–1
Medium sand	0.25–0.5
Fine sand	0.125–0.25
Very fine sand	0.062–0.125
Silt	0.004–0.062
Clay	Less than 0.004

METRIC RULERS

Cut on solid lines.

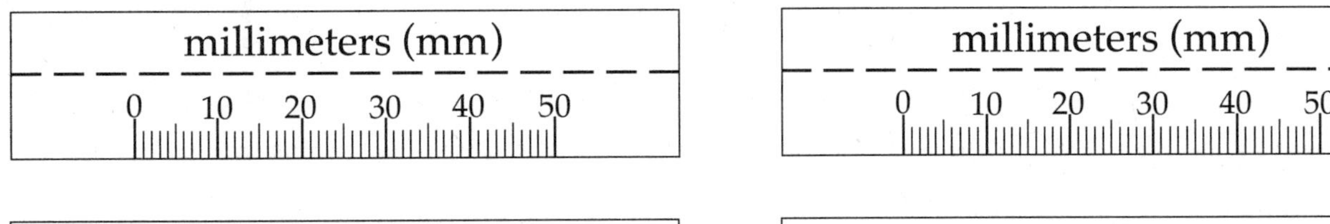

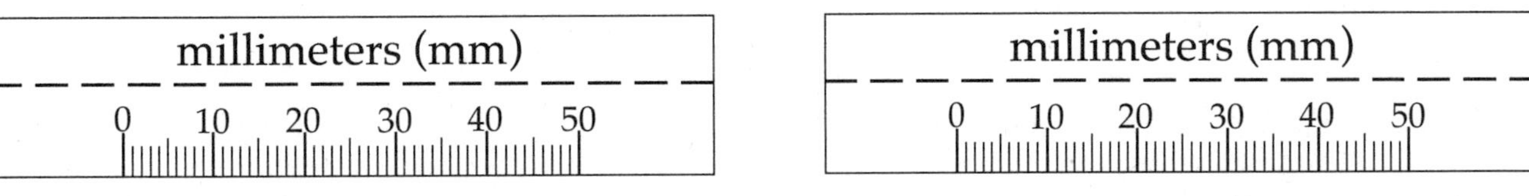

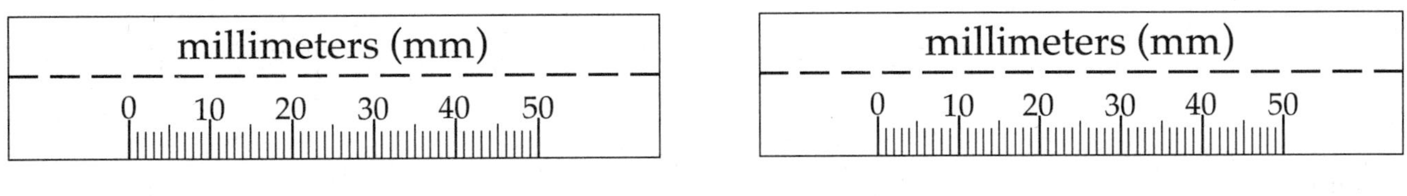

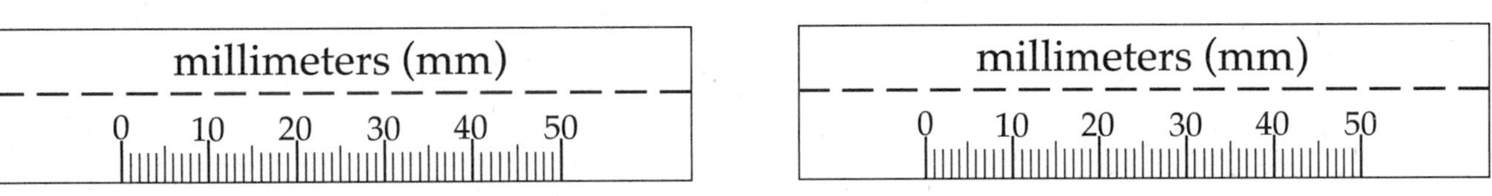

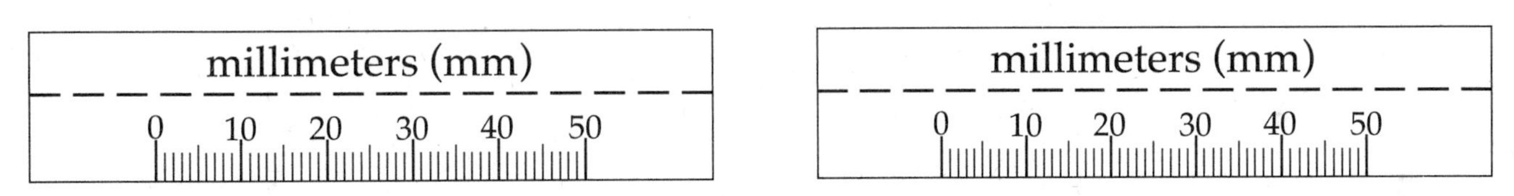

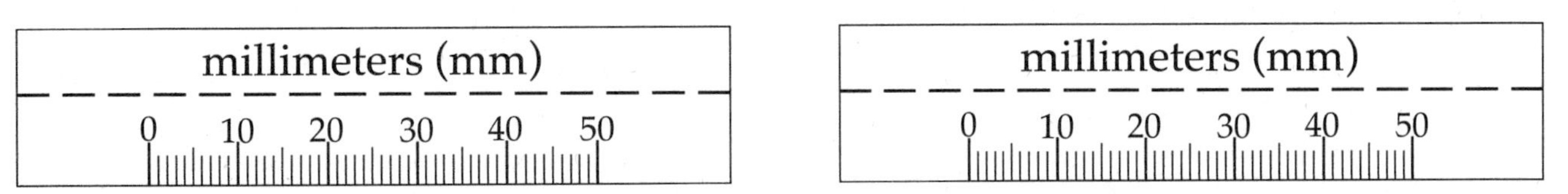

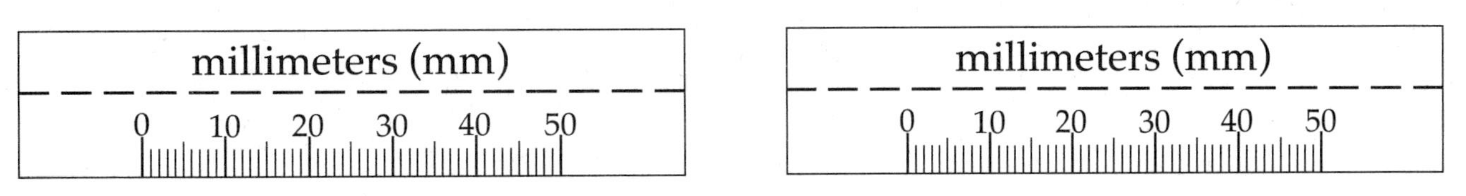

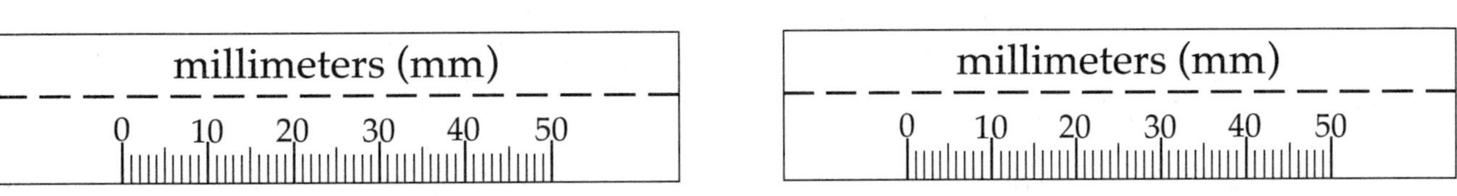

millimeters (mm)
0 10 20 30 40 50

millimeters (mm)
0 10 20 30 40 50

SAND ANALYSIS

Shape. Geologists have names to describe the shape of sand particles.

Angular	Sharp edges; edges are hardly worn off.
Subangular	Slight abrasion; corners and edges are worn off slightly.
Subrounded	Many edges have been noticeably worn off.
Rounded	Edges smoothed; original shape still recognizable.
Well-rounded	Edges and corners totally worn away by abrasion.

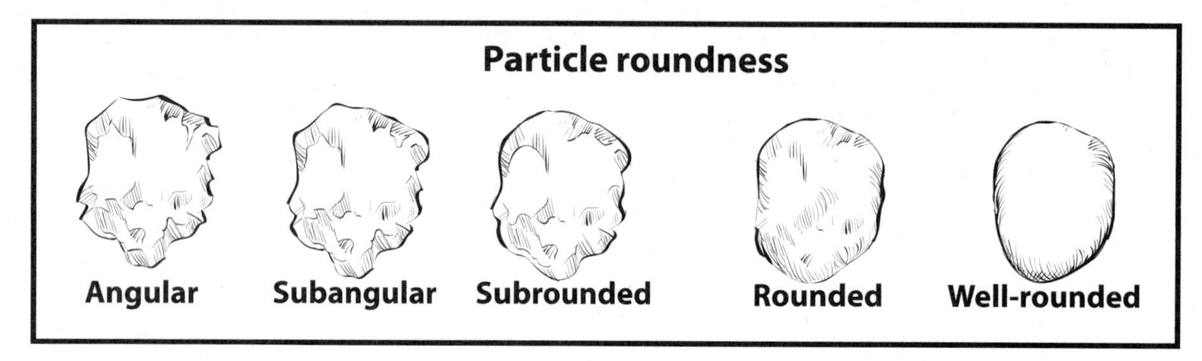

Grain size. Sand is classified into several sizes. See the Wentworth scale for more details about size.

Composition. The composition of sand depends on the rock from which it formed. Knowing the composition of sand can help trace the sand back to its source.

Sorting. Sorting describes how well sand has been separated by size, such as when the stream table carried larger particles farther than smaller particles.

SANDSTONE FORMATION A

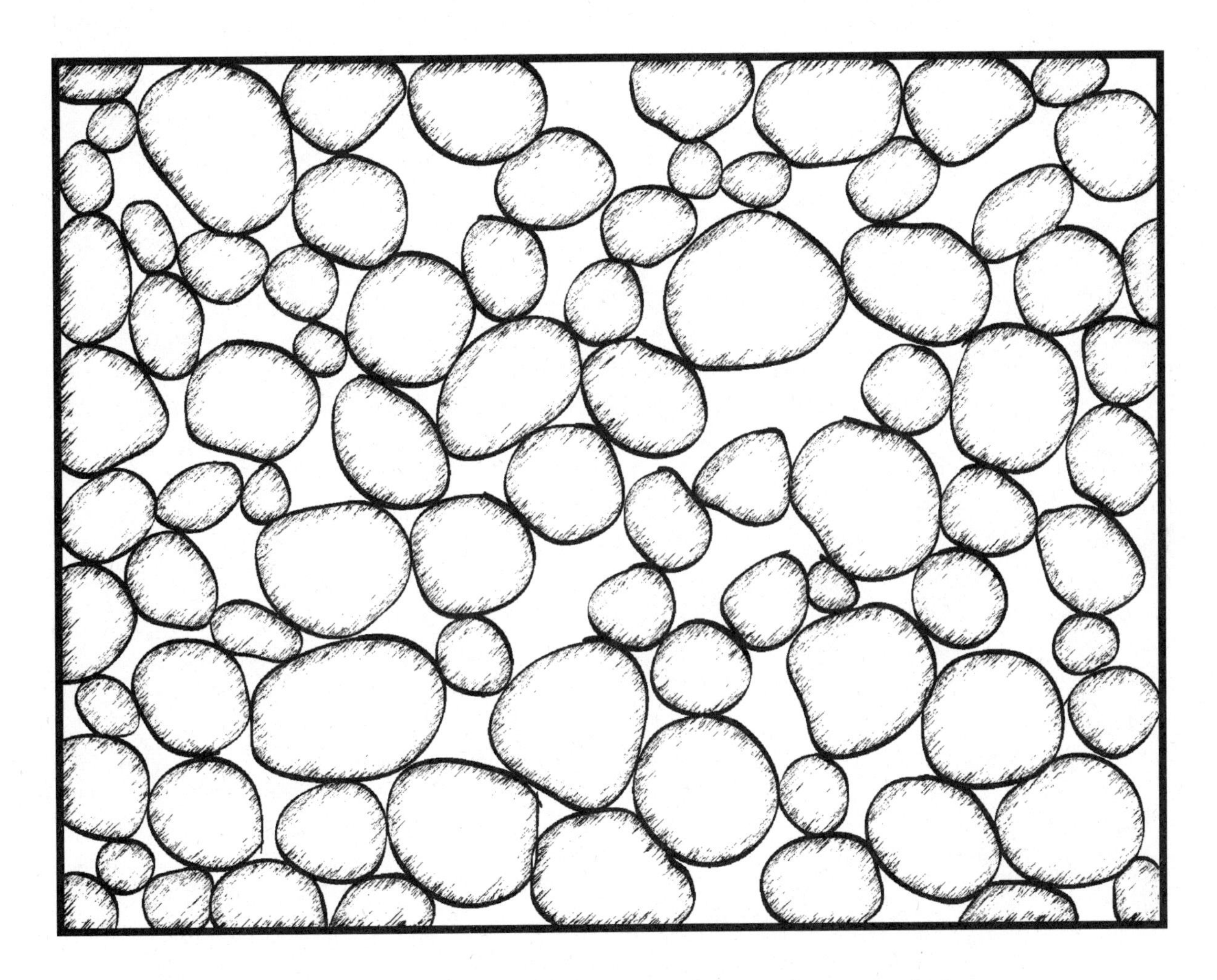

Sand loosely packed soon after deposition

SANDSTONE FORMATION B

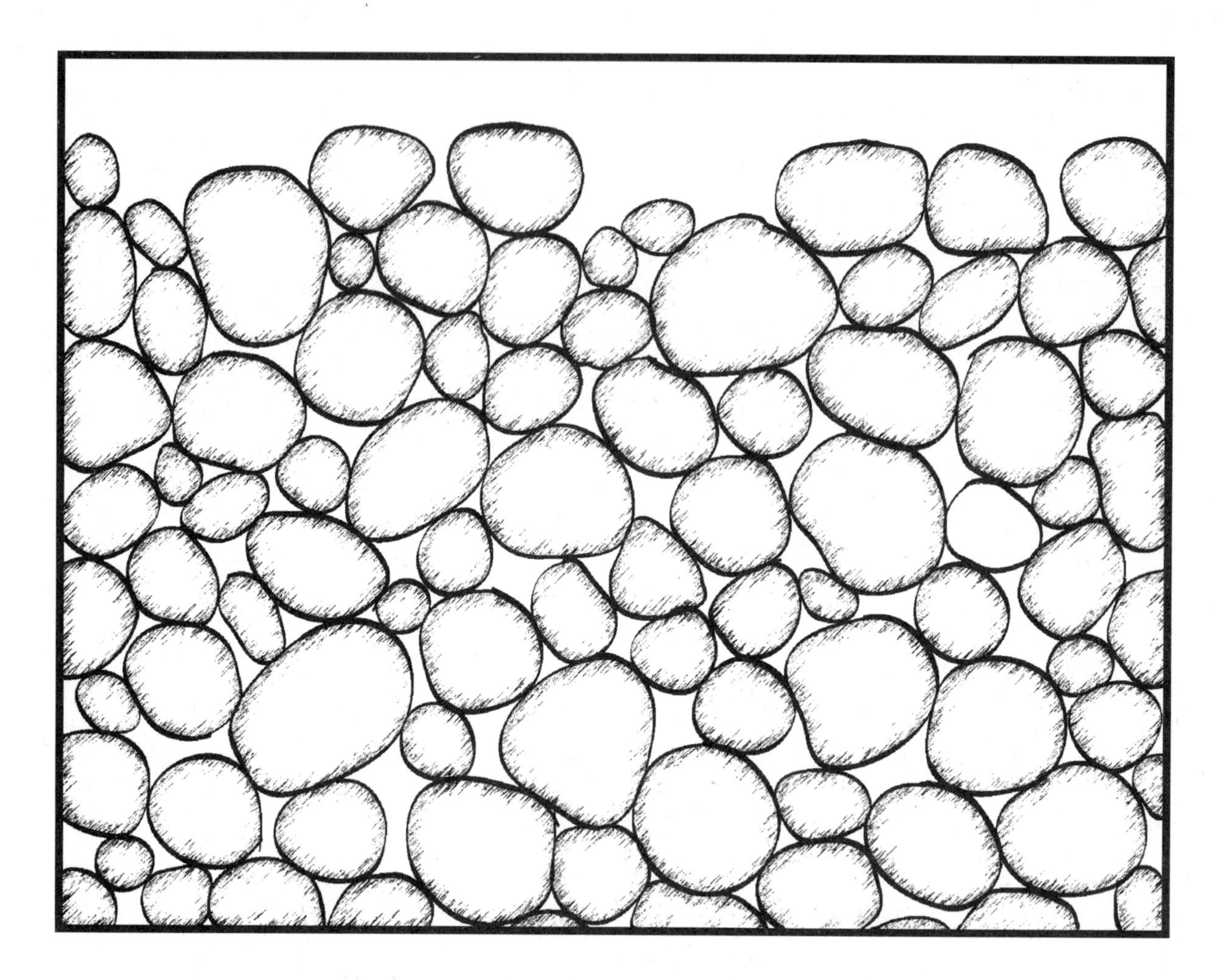

Sand compacted long after deposition

SANDSTONE FORMATION C

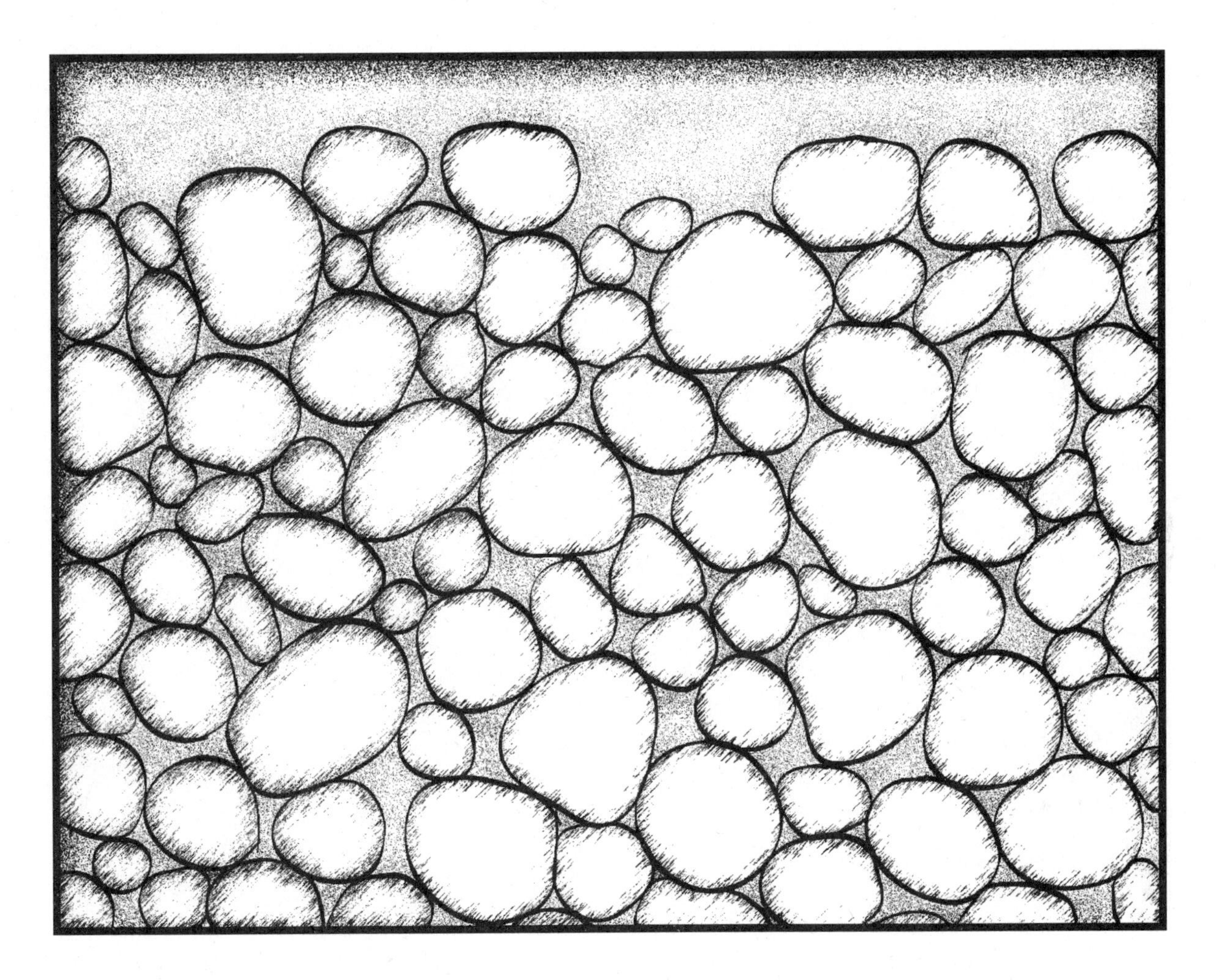

Sand surrounded by cementing mixture

SANDSTONE FORMATION D

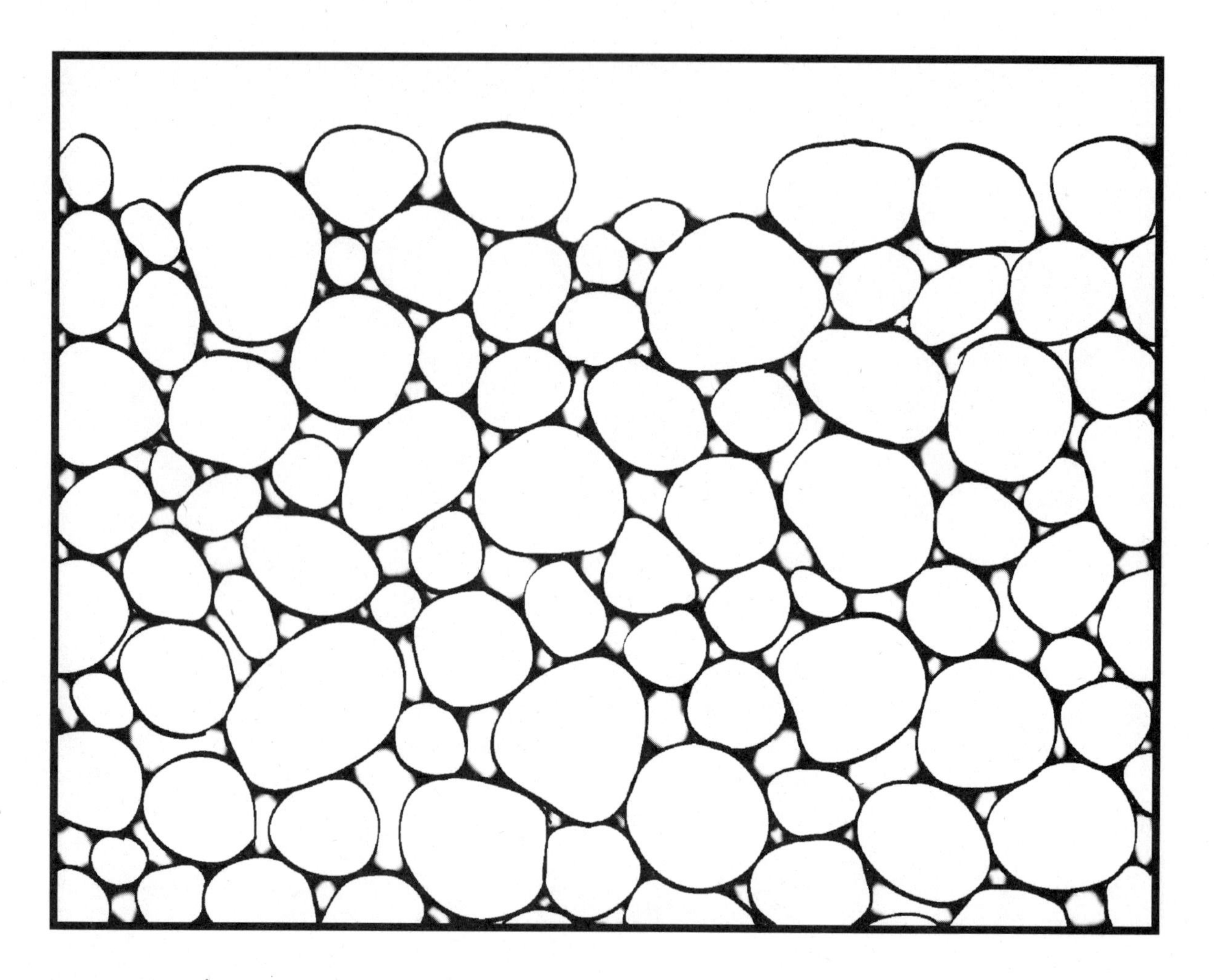

Sandstone

SHALE FORMATION

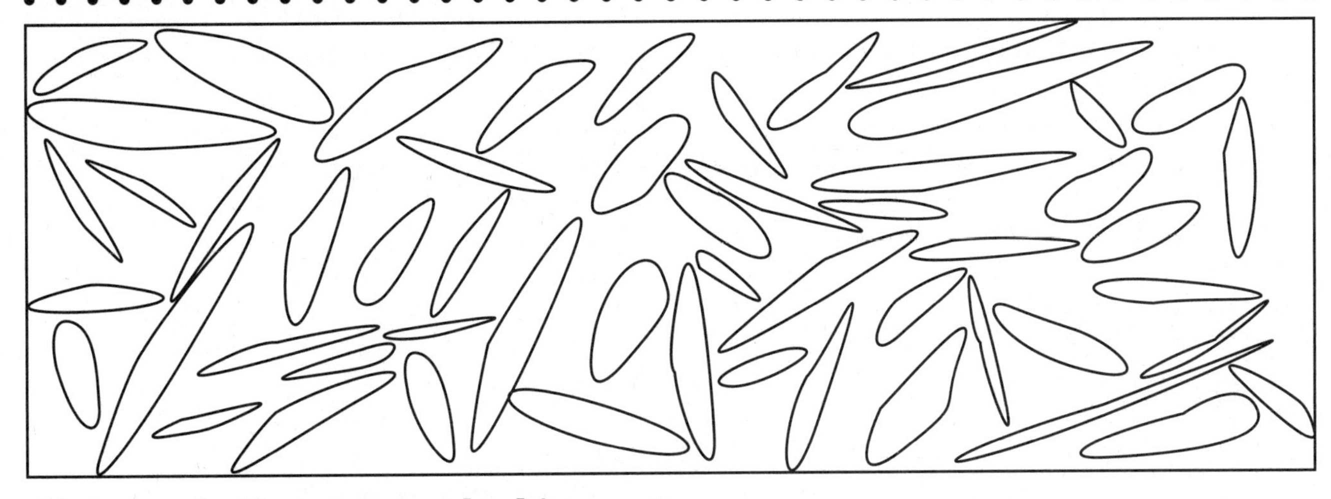

Clay and silt suspended in water

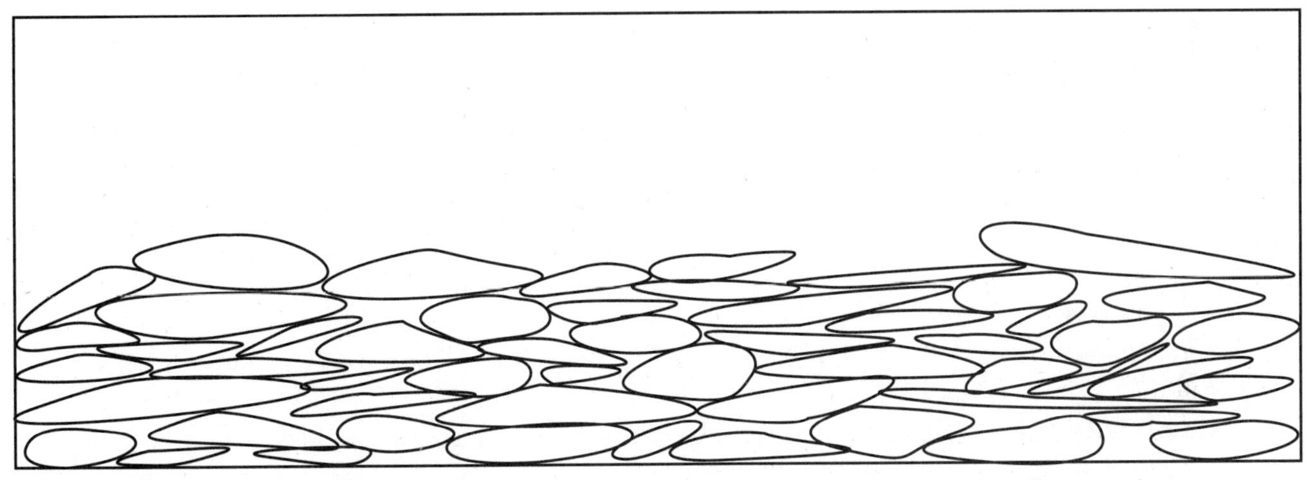

Clay and silt as water is squeezed out during compaction

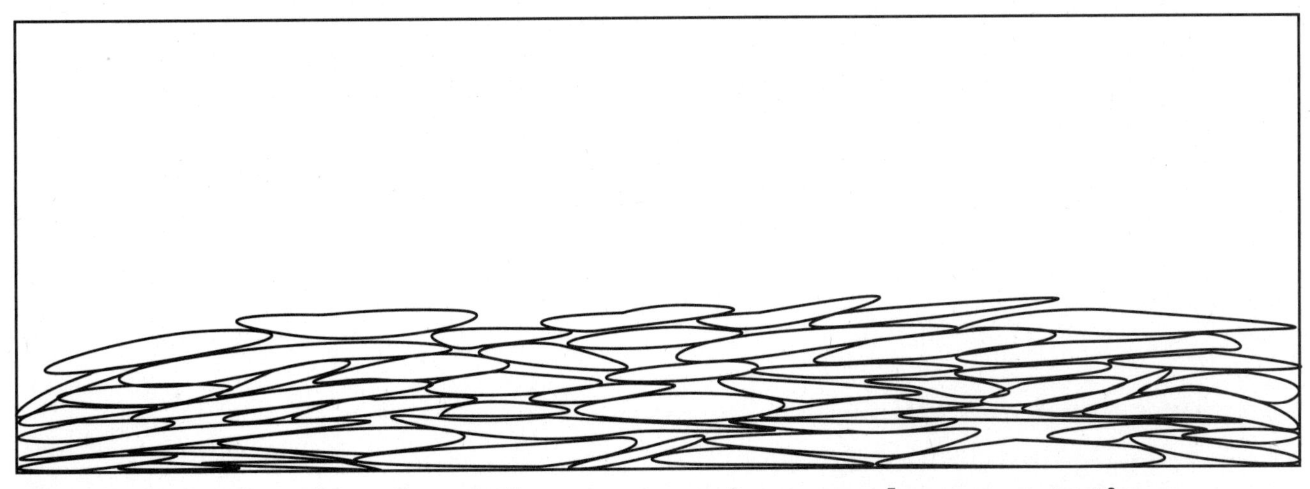

Clay and silt after burial, compaction, and cementation

NOTES ON CALCIUM CARBONATE

- Water is **H_2O**—2 hydrogen atoms bonded to 1 oxygen atom.

- Calcium hydroxide is **$Ca(OH)_2$**—1 calcium atom bonded to 2 oxygen atoms and 2 hydrogen atoms.

- Carbon dioxide is **CO_2**—1 carbon atom bonded to 2 oxygen atoms.

- Calcium carbonate is **$CaCO_3$**—1 calcium atom bonded to 1 carbon atom and 3 oxygen atoms.

- In our investigation, we mixed calcium hydroxide, $Ca(OH)_2$, with carbon dioxide, CO_2. When this happens, the atoms reorganize, a process called a reaction. New products form.

$$\underbrace{Ca + 2O + 2H}_{\substack{Ca(OH)_2 \\ \text{Calcium hydroxide}}} + \underbrace{C + 2O}_{\substack{CO_2 \\ \text{Carbon dioxide}}} \longrightarrow \underbrace{CaCO_3}_{\text{Calcium carbonate}} + \underbrace{H_2O}_{\text{Water}}$$

- The calcium atom bonds with the carbon dioxide plus one of the oxygen atoms from the hydroxide part of the calcium hydroxide to make a molecule of calcium carbonate. The remaining oxygen atom bonds with the two hydrogen atoms to make a molecule of water.

- The calcium carbonate molecule does not dissolve in water; it forms a solid white material.

THE PRINCIPLE OF SUPERPOSITION

- The first layer deposited is the one on the bottom. It is the oldest layer.
- The layers above the first layer are younger as you go up the sequence.

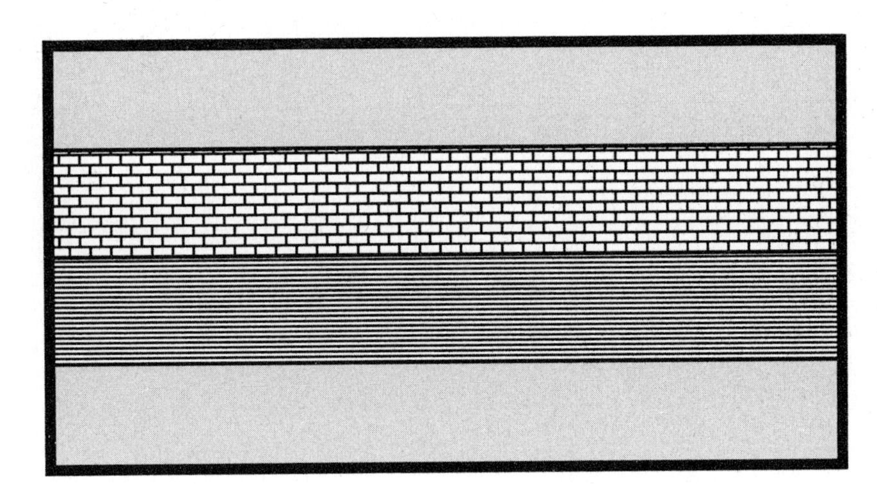

FEATURES OF SEDIMENTARY ROCKS

Ripple marks in Mesnard quartzite

Ripple marks in Hermit shale

Mudcracks in shale, northern Michigan

Ripple marks in Bright Angel shale

Crossbeds in sandstone, Arches National Park

Crossbeds in Navajo sandstone

TIME-LINE SCRIPT A

8. Begin with human history

Remind students that you used a scale of 1 cm for 1 million years. Have students gather around as you point out 1 cm on the time line. Unreel the time line as you add cards. This will keep students together and on task.

Tell them that this time line for Earth starts from modern times, so year 0 is today and is represented by this end of the rope. They will go backward in time on the time line to find out how long ago things happened on Earth. This is like the personal time line, where year 0 was the present and they moved backward in time. Ask,

➤ *Where would your lifetime show up on this scale?* [It is right on the end of the rope. It wouldn't even be seen with a magnifier. It would be 0.001 mm.]

Call up the student holding the *Homo sapiens* card. Ask the student to read the information on the back of the card loud enough for other students to hear, then tell the class where the card should be attached [0.04 cm from the 0 end].

Repeat this process for the following cards. One at a time, each student should read the information and time data aloud, then add the card to the time line.

- *Homo erectus* [1.6 cm]
- Ice age [2.0 cm]
- Primitive humans [3.5 cm]
- First horses [60.0 cm]
- Dinosaur extinction—Cenozoic era begins [66.0 cm]

Point out that each event prior to this point has been during a period of time called the Cenozoic era. Tell students,

*We live in the **Cenozoic** era, which is all of the time since the dinosaurs. **Cenozoic** means "recent life."*

9. Continue with the Mesozoic era

Call up the next round of students, one at a time:

- First birds [1.5 m]
- First true mammals [2.2 m]
- Age of dinosaurs—Mesozoic era begins [2.52 m]

Tell students,

*Geologists call the era from 252 mya to 66 mya the **Mesozoic** era. Mesozoic is derived from the Greek language and means "middle life." Now let's take a look at some of the life that appeared during the Mesozoic.*

Call up the next round of students, one at a time.

- First amphibians [3.6 m]
- First trees [3.85 m]
- First insects [4.0 m]
- First land plants [4.75 m]
- First fish [5.0 m]

TIME-LINE SCRIPT B

Just before the dinosaurs appeared, before the Mesozoic era, there was a great extinction. More than 90% of all species of marine life disappeared at that time. This extinction occurred at the end of what geologists call the ***Paleozoic*** *era, the period of "ancient life." The Paleozoic era lasted from 541 mya to 252 mya.*

Have the student with the card add it to the time line.

- Oldest complex organisms preserved as fossils—Paleozoic era begins [5.4 m]

Tell students,

A number of organisms came into existence during the Paleozoic era. All of the fossils you observed in the Grand Canyon rocks were found in Paleozoic rocks.

10. Continue with the Precambrian era

Tell students,

All of the time before the Paleozoic era is called the ***Precambrian*** *era. It wasn't until the end of the Precambrian era around 600 mya that complex life on Earth evolved, with hard body parts that were easily preserved.*

Call up the last round of students, one at a time.

- First cyanobacteria (blue-green algae)—stromatolites [37 m]
- First life [41 m]
- Earth takes form [46 m]

Have students follow you as the rest of the time line is unfolded, and have students look back to the beginning. Ask,

➤ *Look back along the time line to where we started, to where our lifetime is located. Where do we seem to know the most about Earth's history?* [Close to our lifetimes, recent times.]

➤ *What patterns do you notice in the fossil record when you compare more ancient fossils with more recent fossils?* [More ancient fossils are simple organisms (algae); more recent fossils are more complex organisms like humans, dinosaurs, and birds.]

11. Assess progress: notebook entry

Ask students to write this question in their notebooks and take a few minutes to write a response.

➤ *What main ideas about the history of life on Earth does the time line communicate?*

What to Look For

- *The time line represents the history of life on Earth from the time of its formation, about 4.6 billion years ago (or 46 million years ago).*
- *Events closer to 0 cm are more recent events in history.*
- *Human history is a tiny, recent piece of the vast history of life on Earth, and of the history of Earth itself.*

12. Roll up the time line

Have some students help remove the event cards and roll up the time line. Return to the classroom.

GEOLOGIC TIME SCALE

Era	Period	Epoch	Age (mya)	Origin of Name
Cenozoic (Greek for Life)	Quaternary	**Holocene**	0.01	Greek for wholly recent
		Pleistocene	2.6	
	Tertiary	**Pliocene**	5.3	Greek for most recent
		Miocene	23	Greek for more recent
		Oligocene	34	Greek for less recent
		Eocene	56	Greek for slightly recent
		Paleocene	66	Greek for dawn of recent
Mesozoic (Greek for middle life)	Cretaceous		145	Chalk in southern England
	Jurassic		201	Jura Mountains, Switzerland
	Triassic		252	Rocks in Germany (tri = three)
Paleozoic (Greek for old life)	Permian		299	Province of Perm, Russia
	Pennsylvanian		323	State of Pennsylvania
	Mississippian		359	Mississippi River
	Devonian		419	Devonshire, county in England
	Silurian		444	Silures, Celtic tribe of Wales
	Ordovician		485	Ordovices, Celtic tribe of Wales
	Cambrian		541	Cambria, Roman for Wales
Precambrian			4600	Before Cambrian

Eras. Geological eras include major spans of time based on the life-forms that have been found in rocks of those ages.

Periods. The geological periods are more refined, based on life-forms found in rocks of these ages. It took nearly 100 years to come up with the periods. Their names and time spans are based on rock outcrops in England, Germany, Switzerland, Russia, Wales, and the United States. Some names are based on the geographic areas where the rocks appear at Earth's surface (like Jurassic). Other names are based on the characteristics of the rocks themselves (like Cretaceous).

Epochs. Epochs are subdivisions of the Tertiary and Quaternary periods. English geologist Charles Lyell came up with these subdivisions after he studied marine layers of sedimentary rocks in France and Italy. These names are based on the percentage of fossils in the rocks that are represented by animals and plants still living today. The other periods are also divided into epochs, but those epochs are used mainly by geologists specializing in the rocks that were deposited during those times.

GEOLOGIC TIME LINE A

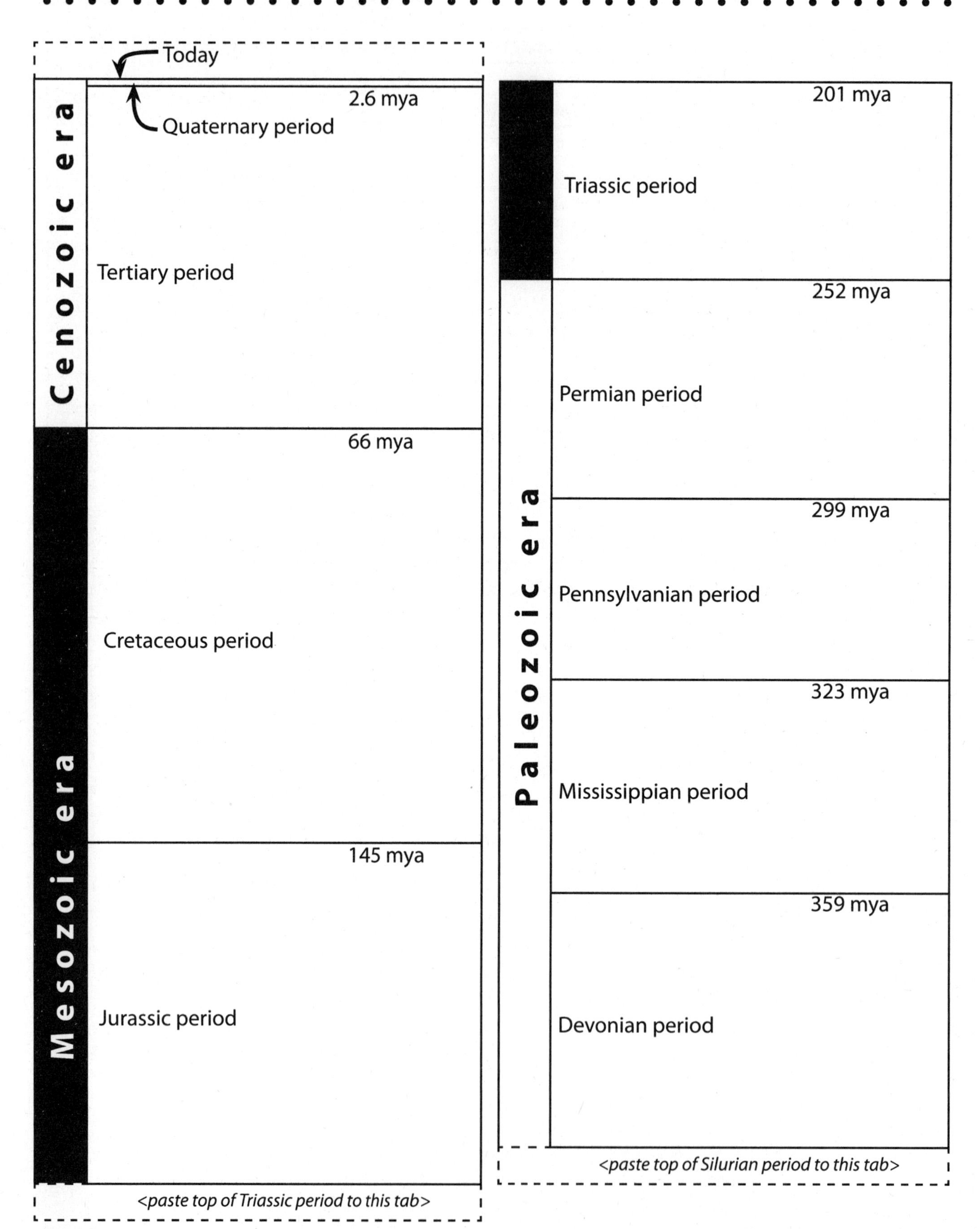

GEOLOGIC TIME LINE B

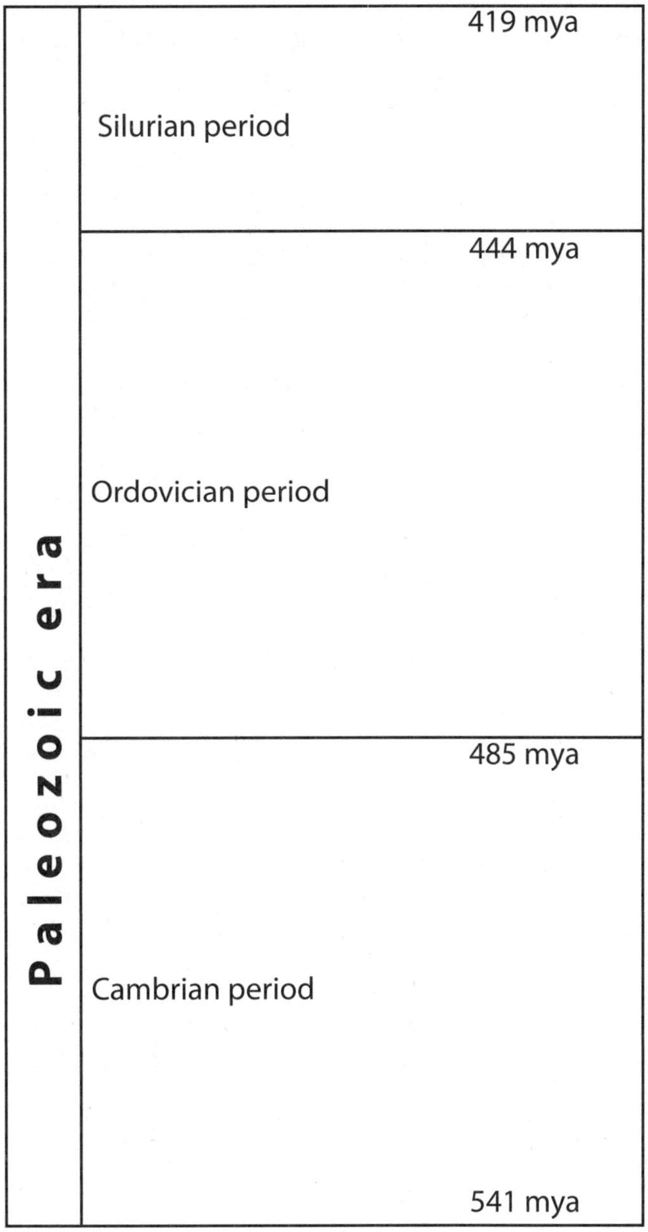

Paleozoic era

419 mya

Silurian period

444 mya

Ordovician period

485 mya

Cambrian period

541 mya

INDEX FOSSILS

Some fossils can be used as index fossils to provide information about the ages of rock layers. Not all fossils are index fossils.

An index fossil must have lived

- for a relatively short period of time;
- in many places around the world.

USES FOR INDEX FOSSILS

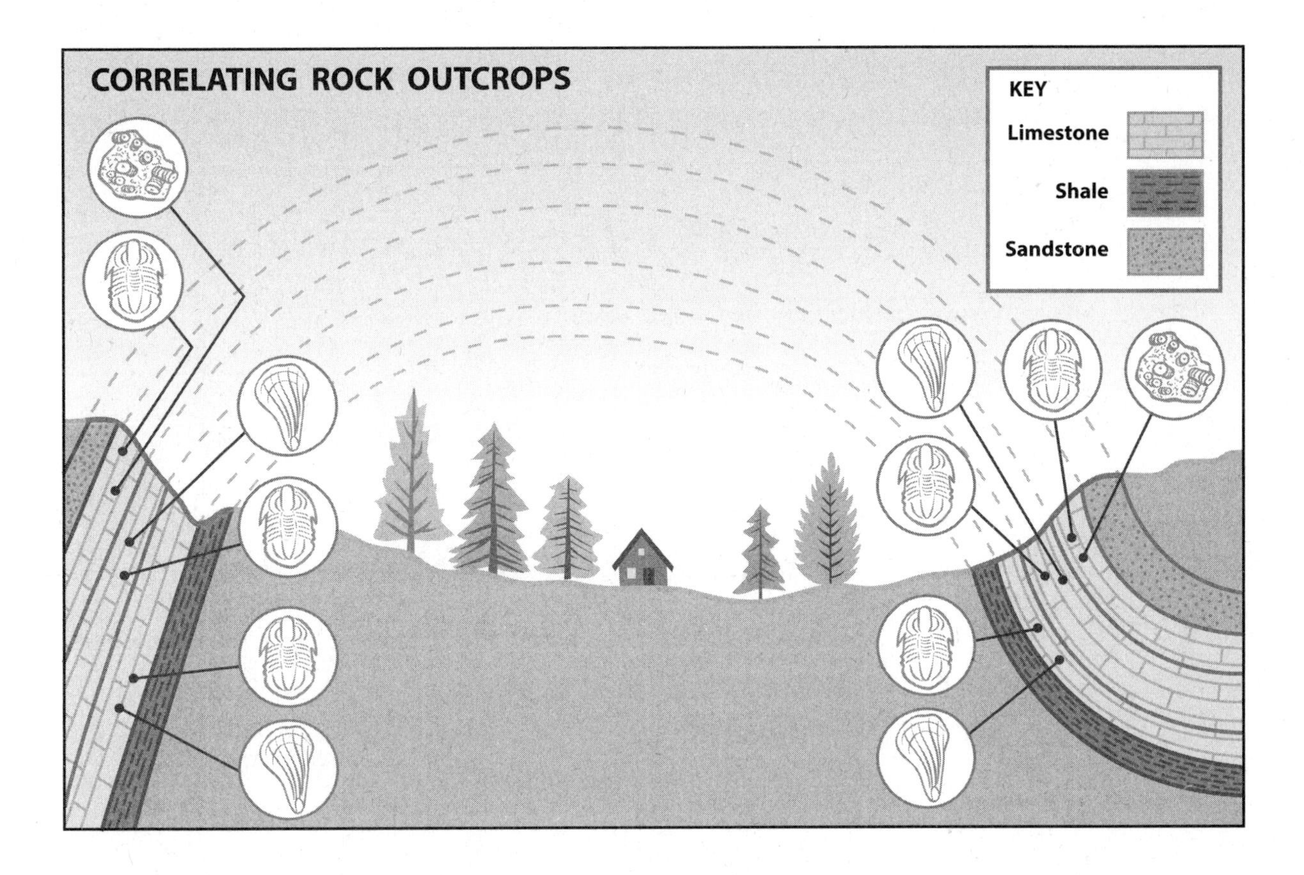

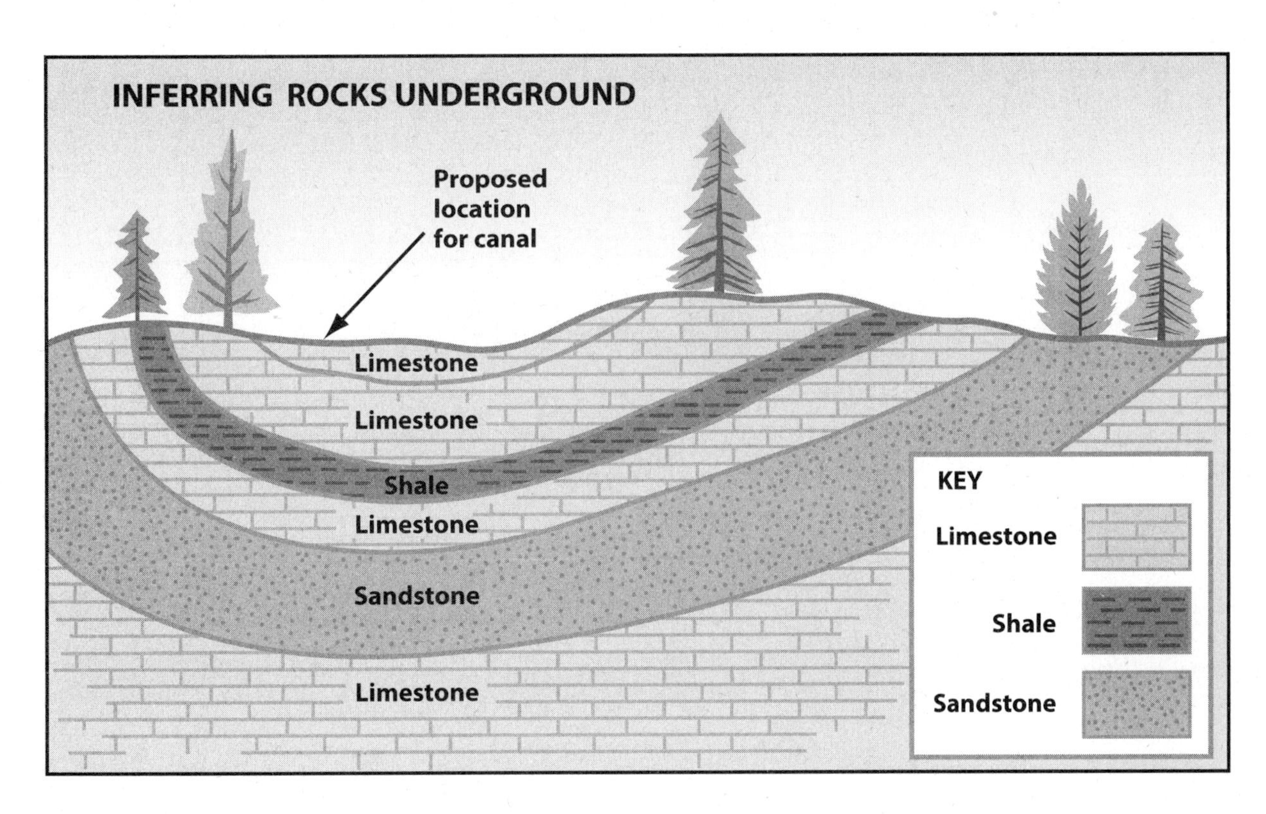

ROCKS OF THE PACIFIC NORTHWEST

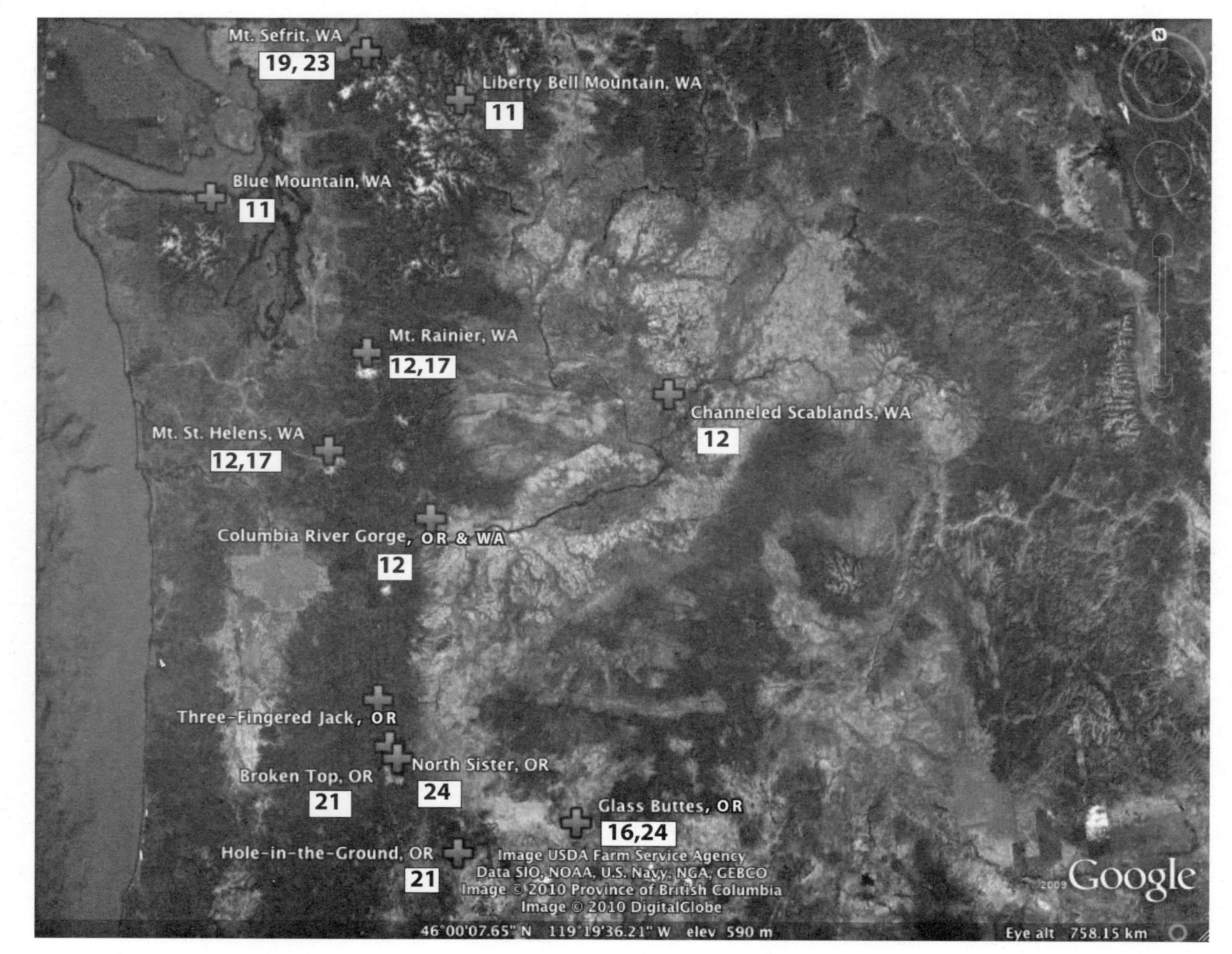

WORLD PLOTTING LOCATIONS

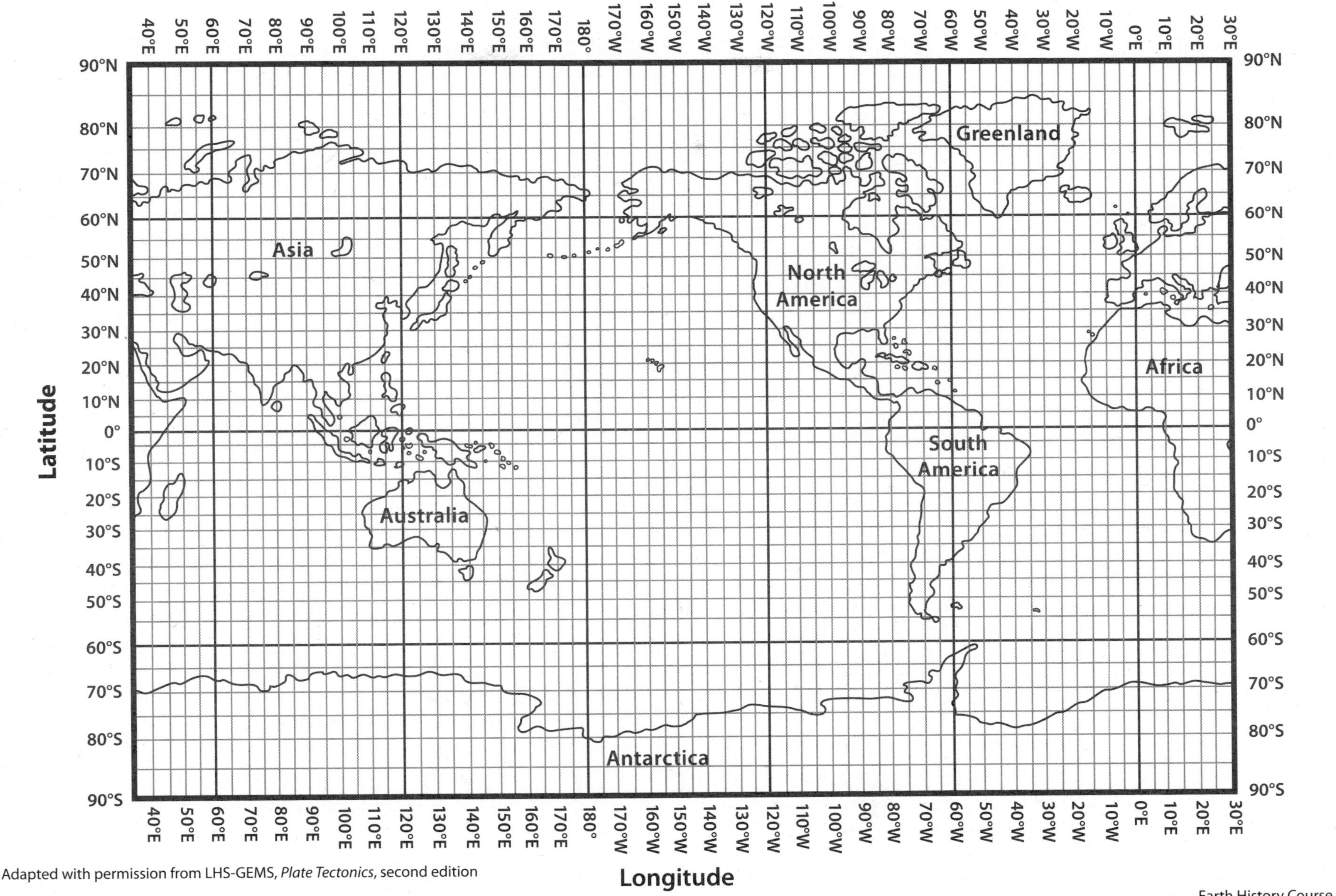

Adapted with permission from LHS-GEMS, *Plate Tectonics*, second edition

US GEOLOGICAL SURVEY PUZZLE

Instructions. Cut out each continent along the edge of the continental shelf (the outermost line).

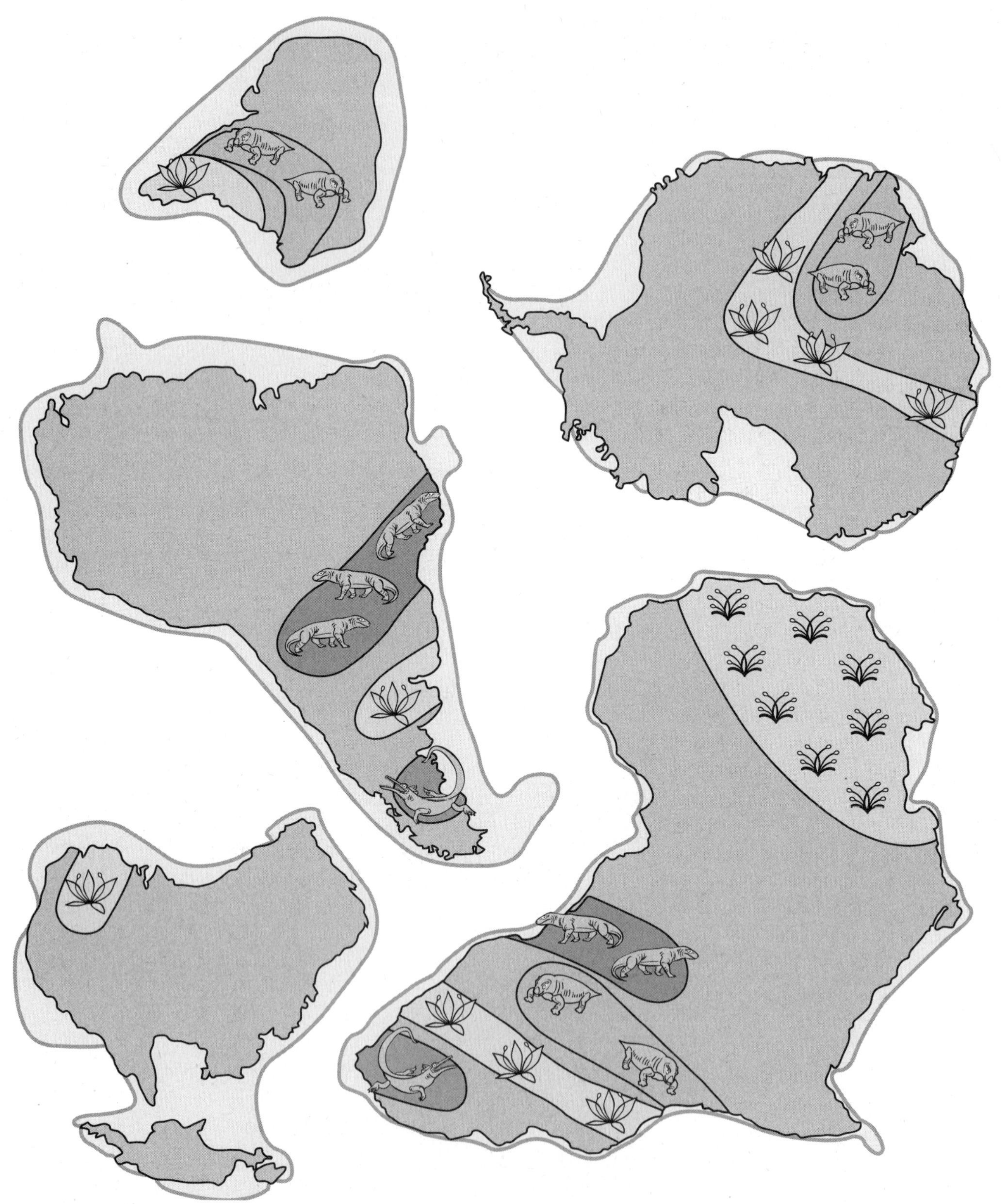

Adapted with permission from *This Dynamic Planet: A Teaching Companion* (US Geological Survey).

US GEOLOGICAL SURVEY PUZZLE KEY

Background

Alfred Wegener's fossil evidence for continental drift is shown on the puzzle pieces. Wegener used this evidence to reconstruct the positions of the continents relative to each other in the distant past.

Instructions

1. Color-code the fossil areas on the puzzle pieces (using the legend below) to make them easier to see.
2. Cut out each of the continents along the edge of the continental shelf (the outermost line).
3. Try to logically piece the continents together so they form a giant supercontinent.
4. When you are satisfied with the fit of the continents, discuss the evidence with your partners.

Key to fossil evidence

By about 300 million years ago (mya), a unique community of plants had evolved, known as the European flora. Fossils of these plants are found in Europe and other areas.

Fossils of the fern *Glossopteris* have been found in these locations.

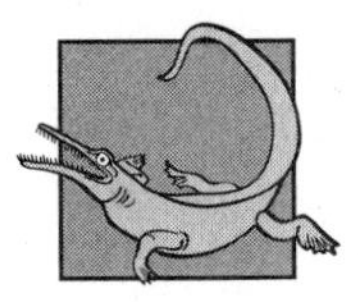

Fossil remains of the half-meter-long aquatic reptile *Mesosaurus* have been found here. Mesosaurs flourished in the early Mesozoic era, about 240 mya. Mesosaurs had limbs for swimming but could also walk on land. Other fossil evidence found in rocks with mesosaurs indicate that they lived in lakes and coastal bays or estuaries.

Fossil remains of *Cynognathus*, a land reptile approximately 3 meters long, were found here. It lived during the early Mesozoic era, about 230 mya. It was a weak swimmer.

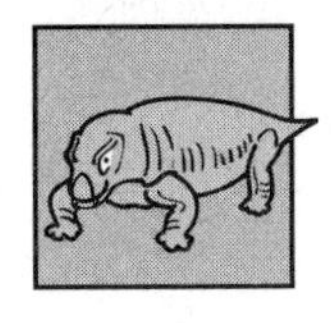

Fossil evidence of the early Mesozoic included the land-dwelling reptile *Lystrosaurus*. It reproduced by laying eggs on land. Its anatomy suggests that this animal was probably a very poor swimmer.

Adapted with permission from *This Dynamic Planet: A Teaching Companion* (US Geological Survey.)

US GEOLOGICAL SURVEY—THE WORLD TODAY

This map shows the continents as they appear today. Most of the continental landmasses lie above sea level, but the true edges of the continents are not at the shoreline. The gray edges of the continents on this map show the relatively shallow water that covers the fringes of the continents. These sea-covered borders are known as continental shelves.

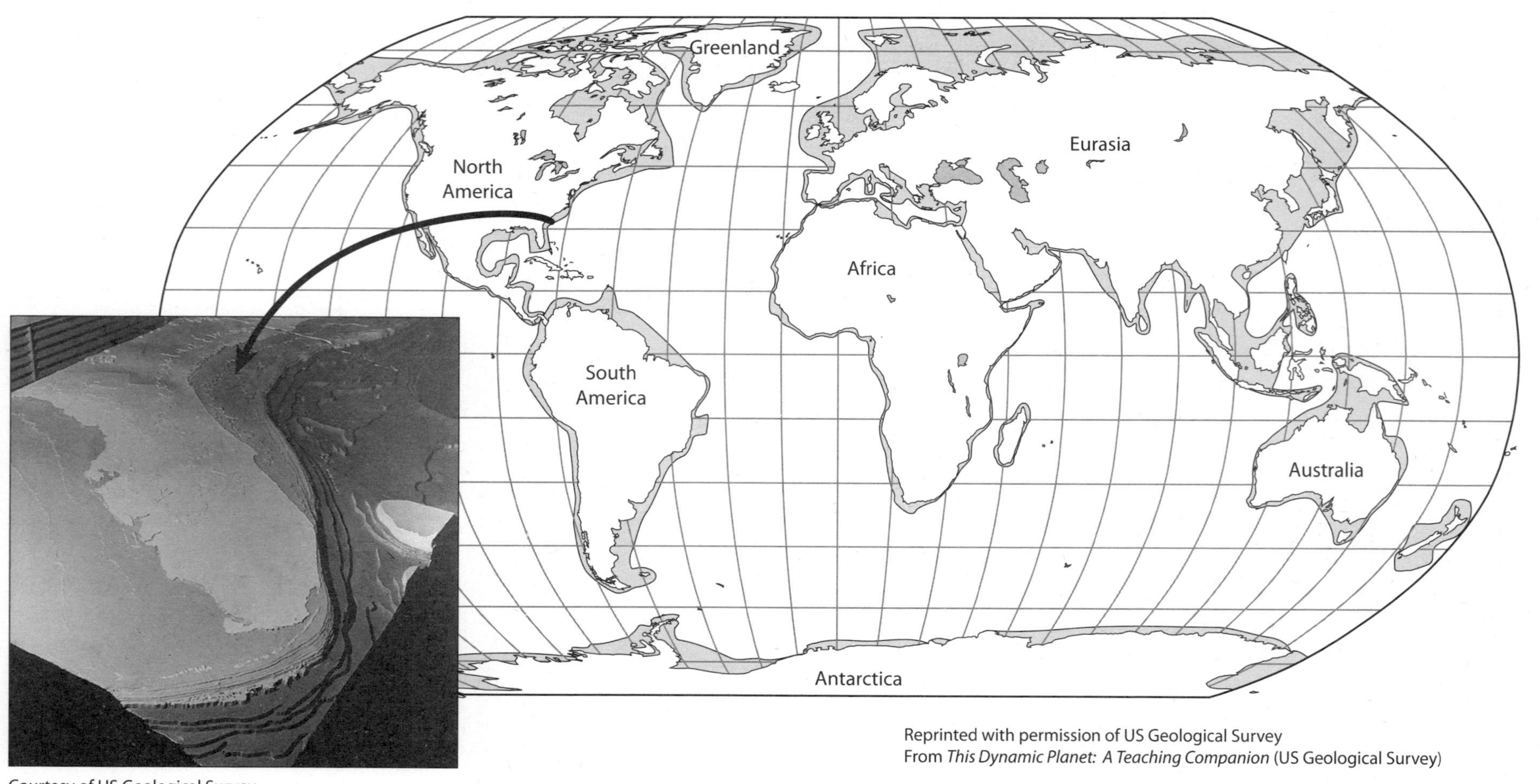

Courtesy of US Geological Survey

A model representing the continental shelf off the Florida coast

Reprinted with permission of US Geological Survey
From *This Dynamic Planet: A Teaching Companion* (US Geological Survey)

PLATE MAP

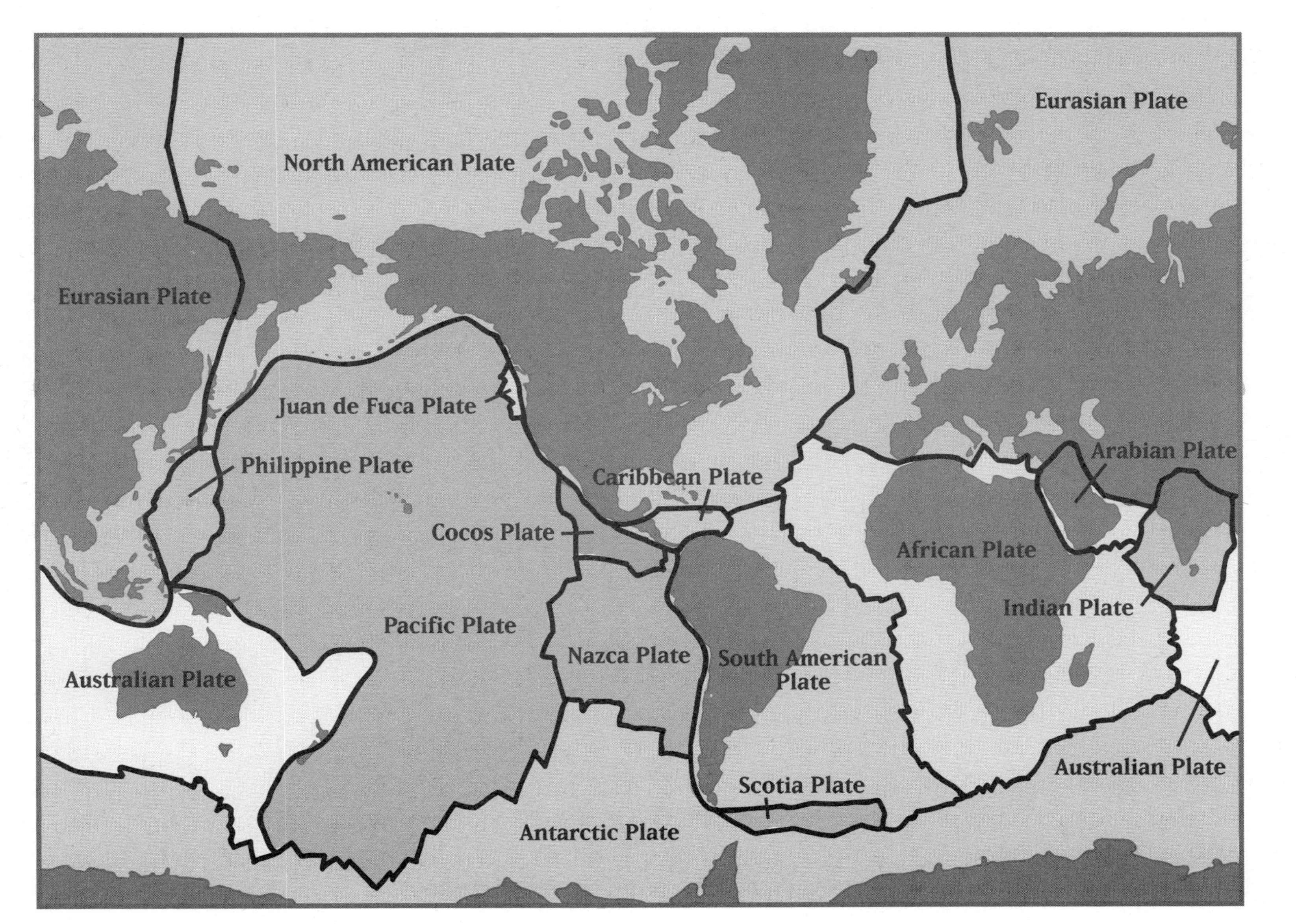

North American Plate
Eurasian Plate
Eurasian Plate
Juan de Fuca Plate
Philippine Plate
Caribbean Plate
Arabian Plate
Cocos Plate
African Plate
Indian Plate
Pacific Plate
Nazca Plate
South American Plate
Australian Plate
Australian Plate
Scotia Plate
Antarctic Plate

EARTH'S LAYERS

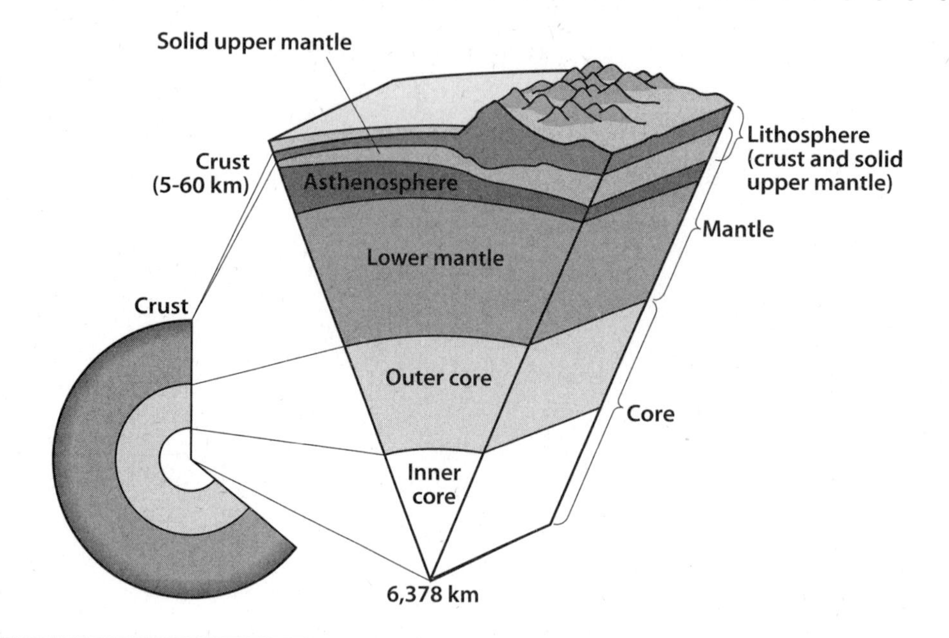

Layer	Consistency	Composition	Temperature	Density	Thickness
Oceanic crust	Rigid, broken into plates	Basalt	From ocean temperature to 870°C near the mantle boundary	More dense than the continental crust	5 km to 8 km
Continental crust	Rigid, broken into plates	Igneous rocks like granite, sedimentary rocks, other rocks	From air temperature to 870°C near the mantle boundary	Less dense than the oceanic crust	30 km to 60 km
Mantle (solid upper mantle)	Rigid, broken into plates	More iron and magnesium than the crust; less silicon and aluminum	500°C to 900°C	More dense than crust and asthenosphere	0 km to 200 km
Mantle (semisolid asthenosphere and lower mantle)	Hot, dense, semisolid	More iron, magnesium, and calcium than the crust	500°C near crust to 4000°C near core	More dense than crust, less dense than solid upper mantle	2,900 km
Outer core	Liquid	Iron, nickel	4400°C to 6100°C	More dense than the mantle	2200 km
Inner core	Solid	Iron, nickel	7000°C	Most dense layer	1,250 km in diameter

PLATE-MOVEMENT MECHANISMS

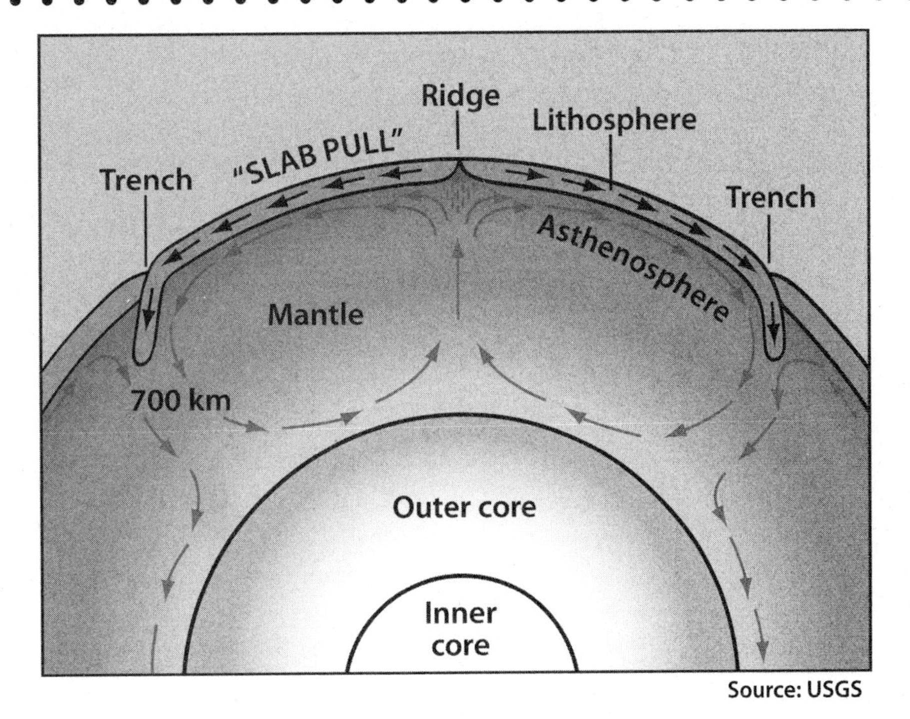

Source: USGS

Near the surface of Earth

- The crust and the hard upper mantle form the lithosphere, which becomes more dense than the underlying asthenosphere as it cools.
- The denser lithosphere could sink into the asthenosphere, moving the lithospheric plate as it sinks.
- The movement of the plates could put stress on the crust, causing it to break apart and move.

Meanwhile, in the mantle

- The asthenosphere is cooler near the top and warmer near the core.
- The cooler part of the asthenosphere could sink toward the core.
- The warmer part of the asthenosphere could rise to the top.
- A convection cycle could be set up in the mantle.

PLATE VOCABULARY

Divergent boundary where plates move away from each other

Convergent boundary where plates move toward each other

Transform boundary where plates move past each another

Convection the rising and falling of fluids, like the mantle, due to heating and cooling

Subduction zone places along a convergent boundary where one plate moves under another plate

Spreading ridge mountain ranges that form on the ocean floor along a divergent boundary

PLATE-MODEL INSTRUCTIONS A

Model A

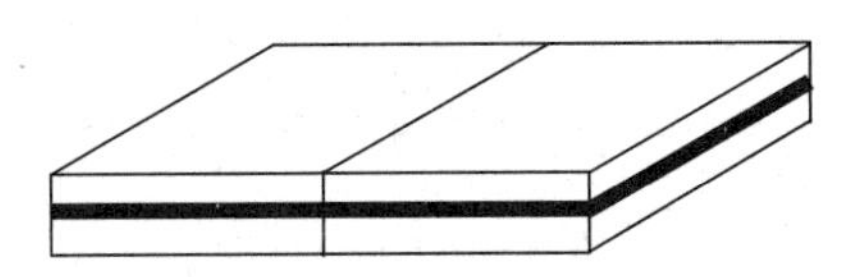

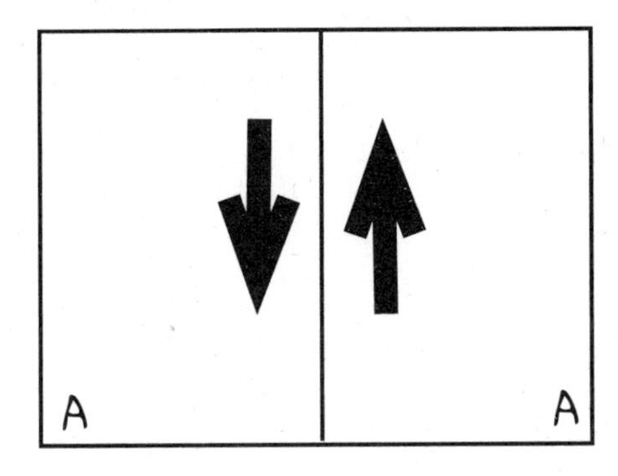

Continental Plate + Continental Plate

Instructions

1. Place the two foam pieces together with the arrows lined up as shown.
2. Slowly push the two pieces together and observe.
3. Slowly pull the two pieces apart and observe.
4. With the pieces touching, slowly slide the two foam pieces past each other in the direction of the arrows. The two sides must touch as you move them.

Questions

1. Draw and describe what you observed in steps 2–4.
2. Where do you see a plate boundary with continental plates pushing together (convergent) on Earth? Pulling apart (divergent)? Sliding past each other (transform)?

Model A

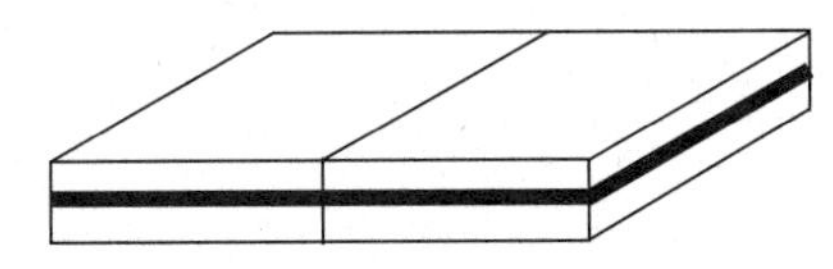

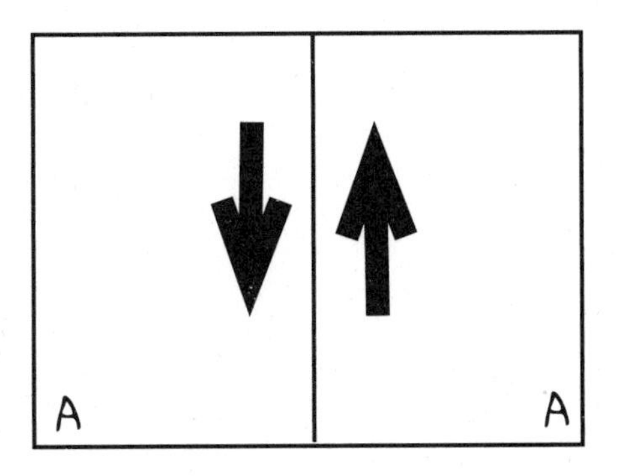

Continental Plate + Continental Plate

Instructions

1. Place the two foam pieces together with the arrows lined up as shown.
2. Slowly push the two pieces together and observe.
3. Slowly pull the two pieces apart and observe.
4. With the pieces touching, slowly slide the two foam pieces past each other in the direction of the arrows. The two sides must touch as you move them.

Questions

1. Draw and describe what you observed in steps 2–4.
2. Where do you see a plate boundary with continental plates pushing together (convergent) on Earth? Pulling apart (divergent)? Sliding past each other (transform)?

PLATE-MODEL INSTRUCTIONS B

Model B

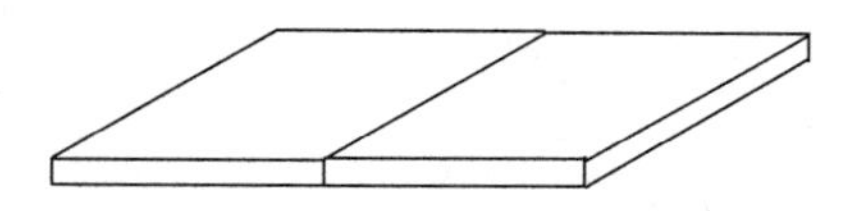

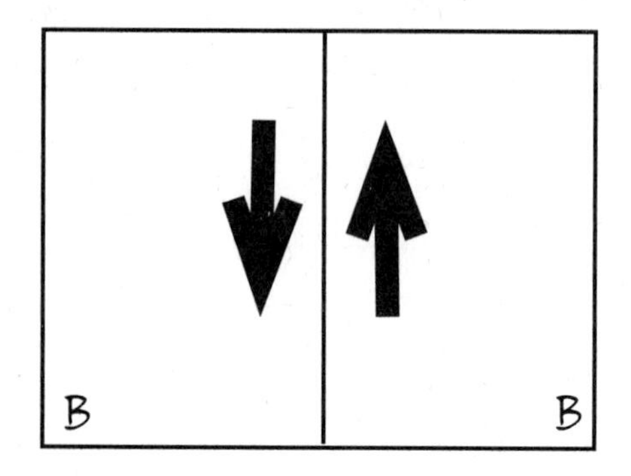

Oceanic Plate + Oceanic Plate

Instructions

1. Place the two cardboard pieces together with the arrows lined up as shown.
2. Slowly push the two pieces together and observe.
3. Slowly pull the two pieces apart and observe.
4. With the pieces touching, slowly slide the two cardboard pieces past each other in the direction of the arrows. The two sides must touch as you move them.

Questions

1. Draw and describe what you observed in steps 2–4.
2. Where do you see a plate boundary with oceanic plates pushing together (convergent) on Earth? Pulling apart (divergent)? Sliding past each other (transform)?

Model B

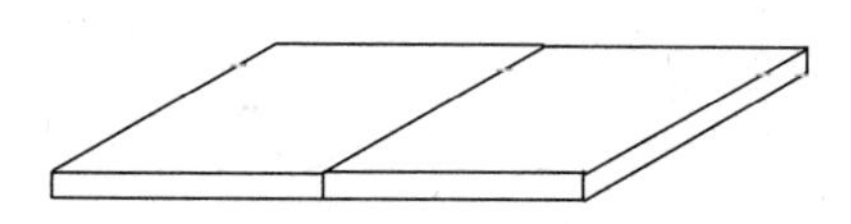

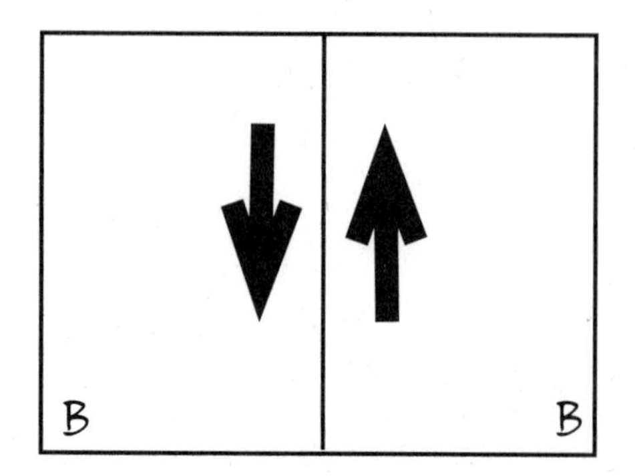

Oceanic Plate + Oceanic Plate

Instructions

1. Place the two cardboard pieces together with the arrows lined up as shown.
2. Slowly push the two pieces together and observe.
3. Slowly pull the two pieces apart and observe.
4. With the pieces touching, slowly slide the two cardboard pieces past each other in the direction of the arrows. The two sides must touch as you move them.

Questions

1. Draw and describe what you observed in steps 2–4.
2. Where do you see a plate boundary with oceanic plates pushing together (convergent) on Earth? Pulling apart (divergent)? Sliding past each other (transform)?

PLATE-MODEL INSTRUCTIONS C

Model C

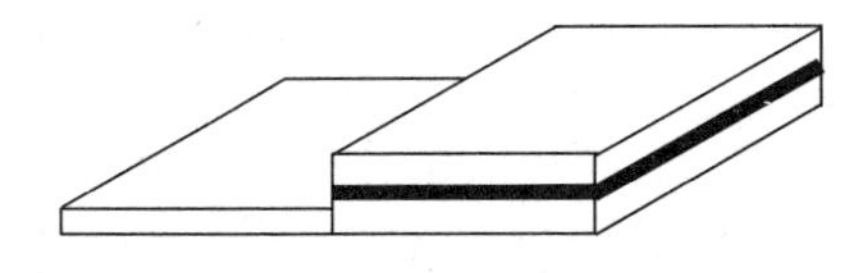

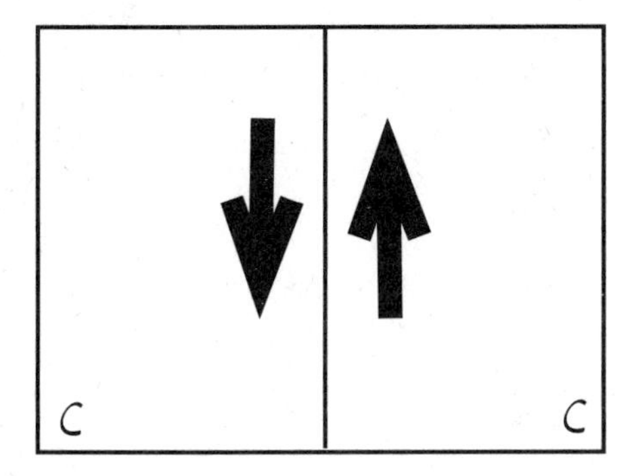

Oceanic Plate + Continental Plate

Instructions

1. Place the foam piece and cardboard piece together with the arrows lined up as shown.
2. Slowly push the two pieces together and observe.
3. Slowly pull the two pieces apart and observe.
4. With the pieces touching, slowly slide the two pieces past each other in the direction of the arrows. The two sides must touch as you move them.

Questions

1. Draw and describe what you observed in steps 2–4.
2. Where do you see a plate boundary with oceanic and continental plates pushing together (convergent) on Earth? Pulling apart (divergent)? Sliding past each other (transform)?

Model C

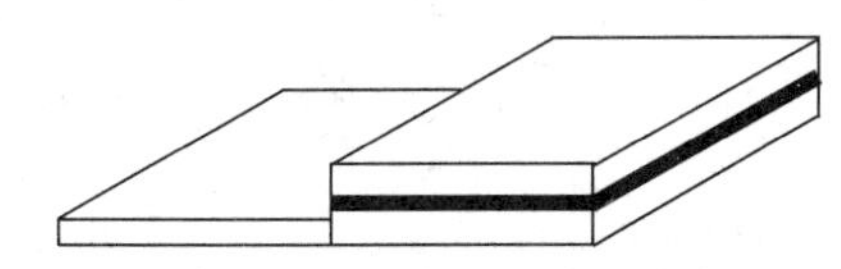

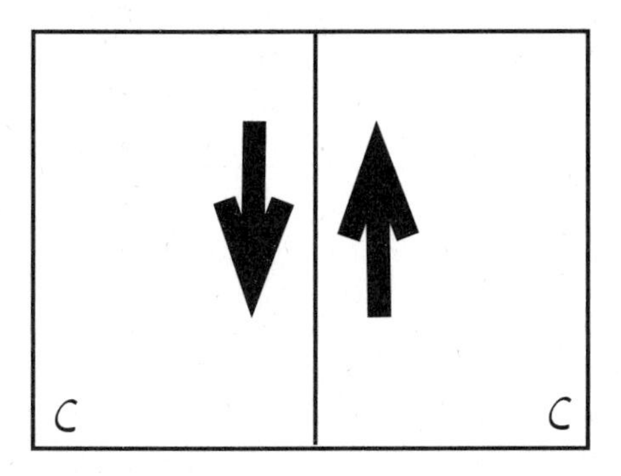

Oceanic Plate + Continental Plate

Instructions

1. Place the foam piece and cardboard piece together with the arrows lined up as shown.
2. Slowly push the two pieces together and observe.
3. Slowly pull the two pieces apart and observe.
4. With the pieces touching, slowly slide the two pieces past each other in the direction of the arrows. The two sides must touch as you move them.

Questions

1. Draw and describe what you observed in steps 2–4.
2. Where do you see a plate boundary with oceanic and continental plates pushing together (convergent) on Earth? Pulling apart (divergent)? Sliding past each other (transform)?

PLATE-MODEL INSTRUCTIONS D

Model D

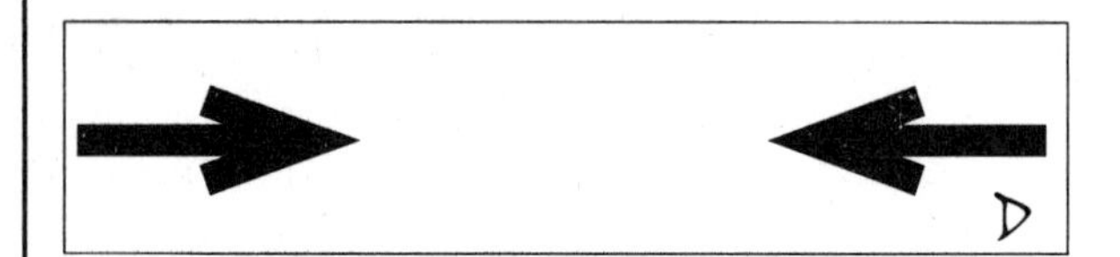

Midcontinent Mountain Building

Instructions

1. Push the two sides of the foam piece together very, very slowly in the direction the arrows point.
2. Observe what happens.

Questions

1. Draw what you observed when you pushed the two sides of the foam together.
2. What did you observe happening to the foam as you pushed the sides together?
3. What would you call the landform that would be created if this happened in Earth's crust?
4. Looking at the plate-interaction map, where do you think landforms like this might occur?

Model D

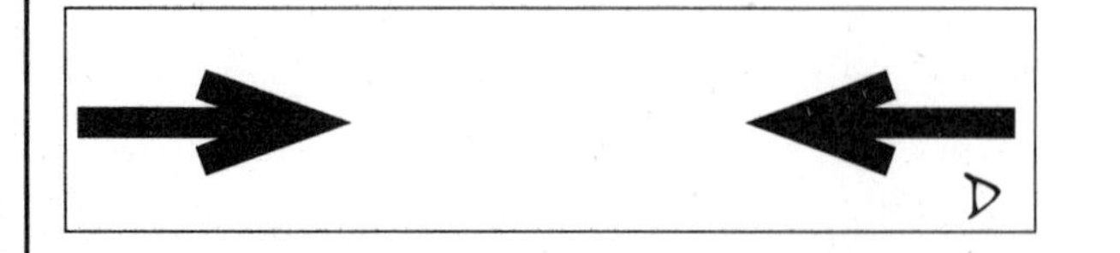

Midcontinent Mountain Building

Instructions

1. Push the two sides of the foam piece together very, very slowly in the direction the arrows point.
2. Observe what happens.

Questions

1. Draw what you observed when you pushed the two sides of the foam together.
2. What did you observe happening to the foam as you pushed the sides together?
3. What would you call the landform that would be created if this happened in Earth's crust?
4. Looking at the plate-interaction map, where do you think landforms like this might occur?

SHENANDOAH STRATIGRAPHIC COLUMN

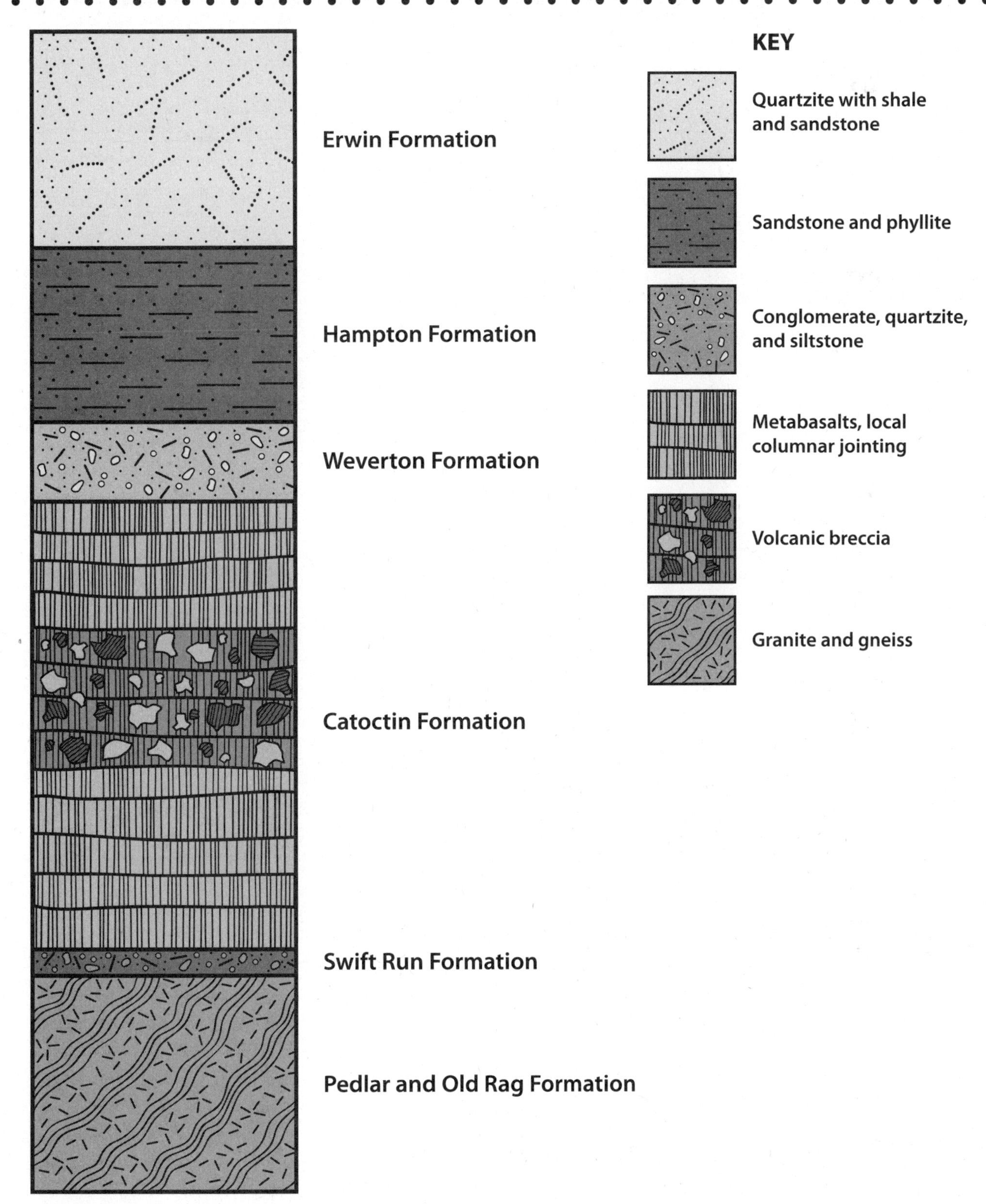

Simplified stratigraphic column for Shenandoah National Park

SKYLINE DRIVE MAP

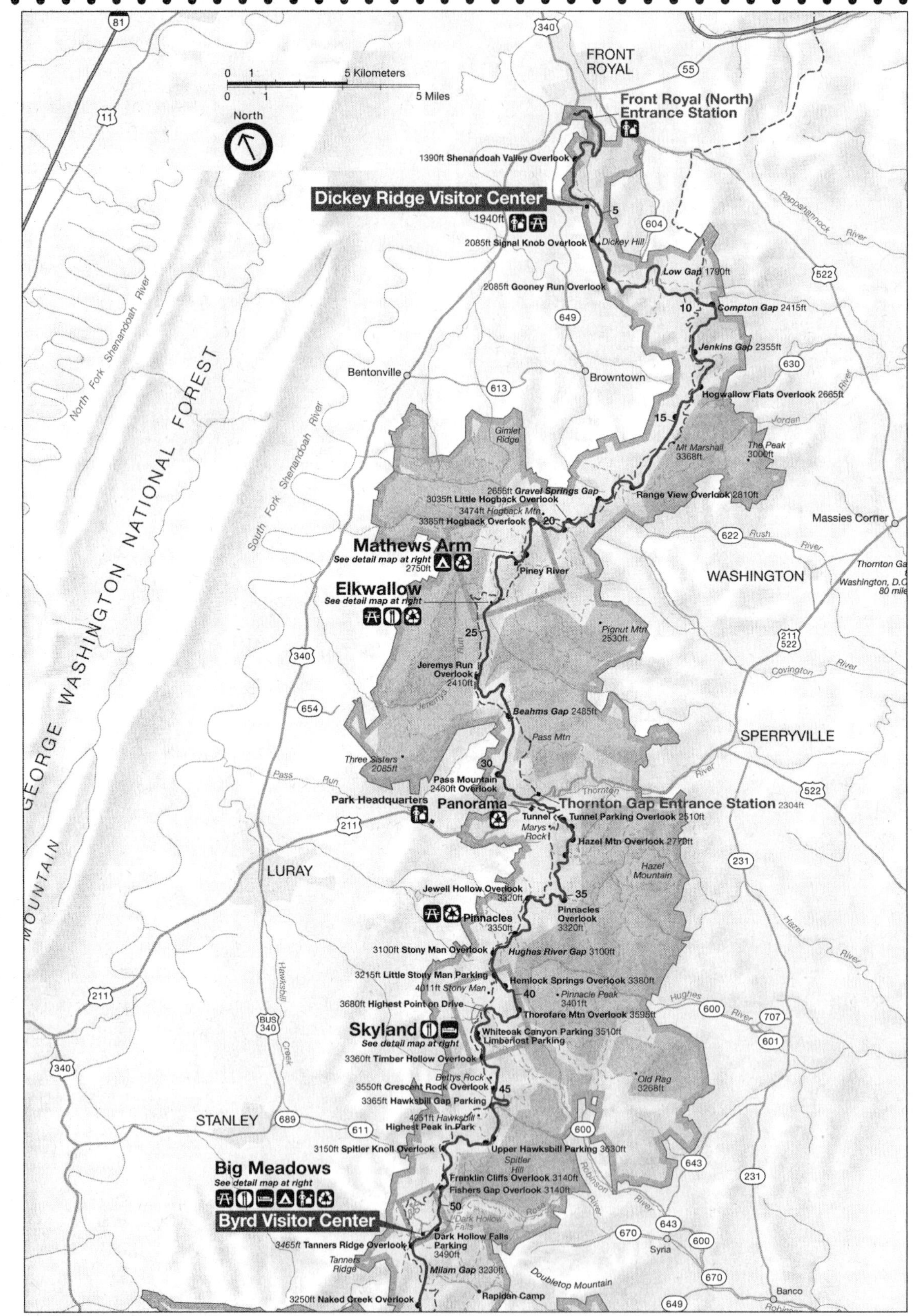

Courtesy of National Park Service

SHENANDOAH TIME LINE

Rock formation or event	Age for time line (mya)
Erosion of Appalachian Mountains continues.	0
Dinosaurs may have roamed in the area. Fossilized footprints have been found east of Shenandoah.	240–65
Opening of the Atlantic Ocean as crust movement reverses.	240
North American and African continents collide. Appalachian Mountains uplifted once again. Extreme metamorphism.	270
First stage of uplift of Appalachian Mountains.	470
Deposition of sediments that would become Weverton, Hampton, and Erwin Formations.	500
First mountain range uplifted, probably the size of present Himalayas. Erosion of early mountain range. Swift Run Formation sediments deposited. Eruption of lava that forms Catoctin Formation.	1,000–570
Intrusion of magma into continental plate and some deposition of sediments. These rocks became the Pedlar and Old Rag Formations.	1,200

GEOSCENARIO GOALS

Each student will

- participate in group work throughout the project, from the introductory tour to the final product and presentation.
- contribute important research from specialist information to the final product and presentation.
- be responsible for explaining all four project questions in the final presentation (described on notebook sheet 45).

Each team must

- create a final presentation about the story of their geoscenario and present it to their peers.
- use available evidence to support their statements.
- create a time line for their geoscenario, including at least one event provided from each team member (which he or she will find in their specialist resource materials).

INTERNET DISCLAIMER

How do you know what to believe on the Internet? Anyone can build a website and put any sort of information on it. The information does not have to be accurate, authenticated, or true. Some websites are designed to deliberately mislead readers about topics, while many others unintentionally contain inaccurate information. Often people reporting news do not have a good understanding of the science behind the story they are covering. So even websites of some national news organizations can contain inaccurate information. Some websites are designed to support the views of a particular political, business, or religious group and may present only information that agrees with their views, even if that viewpoint is not supported by scientific research.

Of course it is important to learn the viewpoints of different groups to know what different people think about how to solve problems. But it is also important to be very aware of who is providing the information, so that you can judge whether the information is accurate or biased. So how do you know what information is accurate and to be trusted as fact? Two questions will help you evaluate websites.

Who published this website?

Some organizations create web content to provide accurate information for the public. Online encyclopedias are a commonly used resource, but an online article can be written by anyone. The author may be a world-renowned expert or someone who is just really interested in the topic. Either way, the author can often include inaccurate information. Online encyclopedias can offer initial broad information on a topic, but they should be only a starting point in your research. You can find footnotes and references at the bottom of each encyclopedia entry, which lead you to more in-depth websites.

Sites that end in **.gov** were created by government organizations and can be a trusted source of information. For accurate geology information, one of the best sources is the US Geological Survey (www.usgs.gov) or the geological survey in your state (www.stategeologists.org). For national parks, check the national parks website (www.nps.gov). Other government sites for scientific information include US Energy Information Administration (www.eia.gov); National Oceanic and Atmospheric Administration (www.noaa.gov); US Environmental Protection Agency (www.epa.gov/students); and Natural Resources Conservation Service (soils.usda.gov).

Sites that end in **.edu** are run by educational organizations such as school districts, colleges, and universities. These are often excellent resources if you can find explanations that aren't too complex.

Sites that end in **.com** are business websites, and sites that end in **.org** are organizations. These sites can offer accurate information, but you should consider their purpose or goal for posting it when judging the validity of information. There are also several organizations aiming to create kid-friendly content that have well-written websites.

What is the website publisher's goal?

Websites don't just present facts. Some websites are designed to persuade readers to take a particular view. For some of the geoscenarios, you will find two lists of websites: those that you can depend on for the most accurate scientific information that is available and those that include the opinions and viewpoints of different groups about a problem or issue related to the geoscenario. The second group may include information not supported by scientific data.

WRITING TIPS

- When using source material that is composed of sentence fragments, transform it into complete sentences that tell part of the story of that place.
- If source material is written in complete sentences, transform it into a labeled diagram, bulleted list, or other format that does NOT use complete sentences, but still contributes to the story.
- When using source material that is a data table, graph, diagram, photo, or drawing, redraw it or write an explanation in your own words to make it better suit the needs of your story.
- Consider clarity: how will you convey the information in your own words in a way that will make sense when you present it to your peers?

TIME LINE OF THE GRAND CANYON

Rock formation or event	Age for time line (mya)
Lava flows dam the Grand Canyon.	3–1
Colorado River begins to erode the Grand Canyon.	6–5
Mountains arise and erode (unconformity).	725
Grand Canyon Supergroup forms.	1,200–740
Zoroaster Granite and other intrusive igneous rocks form.	1,740–1,714
Vishnu Schist and other metamorphic rocks form from existing sedimentary rocks and lava flows.	1,750–1,680
Elves Chasm gneiss forms.	1,840

EARTH HISTORY KEY POINTS

- Every place on Earth's surface has a unique geological story.
- Rocks contain the evidence to make inferences about different environments and events that have occurred on Earth since its origin.
- Many landforms are shaped by slow, persistent processes that proceed over the course of millions of years: weathering, erosion, and deposition.
- Rock can be weathered (broken into smaller pieces) into sediments by a number of processes, including frost wedging, mass wasting (landslides), abrasion (tumbling and scraping), chemical dissolution, and root wedging.
- Sediments can be transported by wind, water, or ice and eventually deposited in low areas or depressions called basins. Some sediments can become widely distributed over Earth's surface as soil.
- Sediments deposited by water usually deposit in flat, horizontal layers (the principle of original horizontality). They can turn into solid rock through the process of lithification.
- The relative ages of sedimentary rock can be determined by the sequence of layers. Lower layers are older than higher layers: this is the principle of superposition.
- The processes that we observe today, for example, weathering, erosion, and deposition, probably happened in the same way over Earth's history. This idea is called uniformitarianism.
- The fossil record represents what we know about ancient life. It is constantly refined as new fossil evidence is discovered.
- Index fossils allow rock layers to be correlated by age, although there are vast distances between the layers.
- Geological time extends from Earth's origin to modern times (today).
- Magma (molten rock under Earth's surface) and lava (molten rock on the surface) can cool. Slower cooling results in larger crystal formation, so crystal size can reveal information about where a rock formed.
- Earth is made of different layers, with the crust being the outermost layer, on which we live.
- Convection currents within the mantle drive plate tectonics and cause plate interactions that result in landforms and processes like earthquakes and volcanism.
- Volcanoes and earthquakes provide clues to the locations of crustal boundaries.
- Plate interactions and movements can form different types of mountains.
- Rocks can be changed into other rocks by weathering, erosion, and deposition; by melting; and by heat and pressure caused by crustal movements or heat from the mantle.
- Metamorphic rocks have changed from the original rock as a result of thermal energy and pressure.

Assessment Masters

Embedded Assessment Notes

Investigation ___, Part ___	Date ___________
Concept:	
Tally: Got it	Doesn't get it
Misconceptions/incomplete ideas:	
Reflections/next steps:	

Investigation ___, Part ___	Date ___________
Concept:	
Tally: Got it	Doesn't get it
Misconceptions/incomplete ideas:	
Reflections/next steps:	

Investigation ___, Part ___	Date ___________
Concept:	
Tally: Got it	Doesn't get it
Misconceptions/incomplete ideas:	
Reflections/next steps:	

Performance Assessment Checklist by Group

	Investigation 1, Part 2		
	Science and Engineering Practices	DCI	Crosscutting Concepts
Group	Planning and carrying out investigations	ESS1.C The history of planet Earth	Patterns

NOTE: See the Assessment chapter for a discussion about how to use this checklist.

Performance Assessment Checklist by Student

	Investigation 1, Part 2		
	Science and Engineering Practices	DCI	Crosscutting Concepts
Student	Planning and carrying out investigations	ESS1.C The history of planet Earth	Patterns

NOTE: See the Assessment chapter for a discussion about how to use this checklist.

Performance Assessment Checklist by Group

	Investigation 3, Part 2				
	Science and Engineering Practices		DCI	Crosscutting Concepts	
Group	Planning and carrying out investigations	Analyzing and interpreting data	ESS2.A Earth materials and systems	Cause and effect	Systems and system models

NOTE: See the Assessment chapter for a discussion about how to use this checklist.

Performance Assessment Checklist by Student

	Investigation 3, Part 2				
	Science and Engineering Practices		DCI	Crosscutting Concepts	
Student	Planning and carrying out investigations	Analyzing and interpreting data	ESS2.A Earth materials and systems	Cause and effect	Systems and system models

NOTE: See the Assessment chapter for a discussion about how to use this checklist.

Performance Assessment Checklist by Group

	Investigation 4, Part 3				
	Science and Engineering Practices		DCI	Crosscutting Concepts	
Group	Analyzing and interpreting data	Constructing explanations	ESS1.C The history of planet Earth	Patterns	Scale, proportion, and quantity

NOTE: See the Assessment chapter for a discussion about how to use this checklist.

Performance Assessment Checklist by Student

	Investigation 4, Part 3				
	Science and Engineering Practices		DCI	Crosscutting Concepts	
Student	Analyzing and interpreting data	Constructing explanations	ESS1.C The history of planet Earth	Patterns	Scale, proportion, and quantity

NOTE: See the Assessment chapter for a discussion about how to use this checklist.

Performance Assessment Checklist by Group

	Investigation 5, Part 1			
	Science and Engineering Practices		DCI	Crosscutting Concepts
Group	Planning and carrying out investigations	Analyzing and interpreting data	ESS2.A Earth materials and systems	Patterns

NOTE: See the Assessment chapter for a discussion about how to use this checklist.

Performance Assessment Checklist by Student

	Investigation 5, Part 1			
	Science and Engineering Practices		DCI	Crosscutting Concepts
Student	Planning and carrying out investigations	Analyzing and interpreting data	ESS2.A Earth materials and systems	Patterns

NOTE: See the Assessment chapter for a discussion about how to use this checklist.

Performance Assessment Checklist by Group

	Investigation 5, Part 2					
	Science and Engineering Practices			DCI	Crosscutting Concepts	
Group	Planning and carrying out investigations	Analyzing and interpreting data	Constructing explanations	ESS2.A Earth materials and systems	Patterns	Cause and effect

NOTE: See the Assessment chapter for a discussion about how to use this checklist.

Performance Assessment Checklist by Student

	Investigation 5, Part 2					
	Science and Engineering Practices			DCI	Crosscutting Concepts	
Student	Planning and carrying out investigations	Analyzing and interpreting data	Constructing explanations	ESS2.A Earth materials and systems	Patterns	Cause and effect

NOTE: See the Assessment chapter for a discussion about how to use this checklist.

Performance Assessment Checklist by Group

	Investigation 6, Part 1			
	Science and Engineering Practices		DCI	Crosscutting Concepts
Group	Analyzing and interpreting data	Using mathematics and computational thinking	ESS3.B Natural hazards	Patterns

NOTE: See the Assessment chapter for a discussion about how to use this checklist.

Performance Assessment Checklist by Student

	Investigation 6, Part 1			
	Science and Engineering Practices		DCI	Crosscutting Concepts
Student	Analyzing and interpreting data	Using mathematics and computational thinking	ESS3.B Natural hazards	Patterns

NOTE: See the Assessment chapter for a discussion about how to use this checklist.

Performance Assessment Checklist by Group

	Investigation 7, Part 2				
	Science and Engineering Practices		DCI	Crosscutting Concepts	
Group	Analyzing and interpreting data	Engaging in argument from evidence	ESS2.A Earth materials and systems	Patterns	Cause and effect

NOTE: See the Assessment chapter for a discussion about how to use this checklist.

Performance Assessment Checklist by Student

	Investigation 7, Part 2				
	Science and Engineering Practices		DCI	Crosscutting Concepts	
Student	Analyzing and interpreting data	Engaging in argument from evidence	ESS2.A Earth materials and systems	Patterns	Cause and effect

NOTE: See the Assessment chapter for a discussion about how to use this checklist.

Performance Assessment Checklist by Group

	Investigation 8, Part 3, Page 1 of 2		
	Science and Engineering Practices		
Group	Asking questions	Engaging in argument from evidence	Obtaining, evaluating, and communicating information

NOTE: See the Assessment chapter for a discussion about how to use this checklist.

Performance Assessment Checklist by Student

	Investigation 8, Part 3, Page 1 of 2		
	Science and Engineering Practices		
Student	Asking questions	Engaging in argument from evidence	Obtaining, evaluating, and communicating information

NOTE: See the Assessment chapter for a discussion about how to use this checklist.

Performance Assessment Checklist by Group

	Investigation 8, Part 3, Page 2 of 2			
	Disciplinary Core Ideas		Crosscutting Concepts	
Group	ESS3.A Natural resources	ESS3.C Human impact on Earth systems	Cause and effect	Stability and change

NOTE: See the Assessment chapter for a discussion about how to use this checklist.

Performance Assessment Checklist by Student

	Investigation 8, Part 3, Page 2 of 2			
	Disciplinary Core Ideas		Crosscutting Concepts	
Student	ESS3.A Natural resources	ESS3.C Human impact on Earth systems	Cause and effect	Stability and change

NOTE: See the Assessment chapter for a discussion about how to use this checklist.

Assessment Record—Entry-Level Survey

Item	Contributes to	Notes for planning instruction
1a	MS-ESS2-2	
1b	MS-ESS3-4 MS-ESS3-5	
2	MS-ESS1-4 MS-ESS2-2 MS-LS4-1	
3	MS-ESS2-1 MS-ESS2-3	
4	MS-ESS3-2	
5	MS-ESS3-1	
6	MS-ESS1-4 MS-ESS2-2 MS-LS4-1	

Assessment Record—Investigations 1–2 I-Check

Student names	1	2	3a	3b	4	5	6	7a	7b

NOTE: A spreadsheet for this chart is available on www.FOSSweb.com.

Assessment Record—Investigations 3–4 I-Check

Student names	1a	1b	1c	2a	2b	3	4	5	6	7a	7b	8a	8b

NOTE: A spreadsheet for this chart is available on www.FOSSweb.com.

Assessment Record—Investigations 5–6 I-Check

Student names	1	2	3	4a	4b	5	6	7	8	9

NOTE: A spreadsheet for this chart is available on www.FOSSweb.com.

Assessment Record—Investigation 7 I-Check

Student names	1a	1b	2a	2b	3a	3b	3c	4a	4b	5

NOTE: A spreadsheet for this chart is available on www.FOSSweb.com.

Assessment Record—Posttest, Page 1 of 2

Student names	1a	1b	2	3a	3b	4a	4b	4c

NOTE: A spreadsheet for this chart is available on www.FOSSweb.com.

Assessment Record—Posttest, Page 2 of 2

Student names	5a	5b	5c	6	7	8	9	10a	10b

NOTE: A spreadsheet for this chart is available on www.FOSSweb.com.

Name ______________________________

ENTRY-LEVEL SURVEY
EARTH HISTORY

Date ___________ Class ___________

1. Earth's major systems are
   - the geosphere (solid and molten rock, soil, and sediments)
   - the hydrosphere (water)
   - the atmosphere (air)
   - the biosphere (living things, including humans)

   a. Earth's major systems interact in many ways to change Earth's surface. Describe two examples of changes caused by the interaction of the geosphere and one of the other systems listed above.

   ____

   ____

   ____

   ____

   ____

   ____

   b. Describe two examples of how human activity has affected Earth's systems and how people are trying to reduce the effects of those activities.

   ____

   ____

   ____

   ____

   ____

   ____

   ____

   ____

   ____

   ____

Name ______________________

ENTRY-LEVEL SURVEY
EARTH HISTORY

2. Student A says that "Earth has changed over a very long period of time." Student B says, "What evidence do scientists have to support that argument?"

   If you were student A, how would you answer student B's question?

   ____
   ____
   ____
   ____
   ____
   ____
   ____
   ____
   ____

   Student C adds, "Actually, Earth has always changed and continues to change in the same ways." Do you agree with student C? Why or why not?

   ____
   ____
   ____
   ____
   ____
   ____
   ____
   ____
   ____

Name ______________________________

ENTRY-LEVEL SURVEY

EARTH HISTORY

3. How does earth material cycle within the geosphere? As part of your explanation, consider how energy from inside Earth plays a role.

____

____

____

____

____

____

____

____

____

____

____

4. A variety of hazards (like earthquakes, volcanic eruptions, etc.) result from natural geological processes. Provide examples of how humans can reduce the impact of natural hazards.

____

____

____

____

____

____

____

____

____

Name ______________________

ENTRY-LEVEL SURVEY

EARTH HISTORY

5. Consider the following examples of resources humans use. Identify if each resource is a renewable (R) or nonrenewable (N) resource and explain why.

Resource	R / N	Explanation
The Sun		
Fossil fuels		
Water		
Geothermal power		
Soil		
Minerals		
Air		
Trees		

Name ___________________________

ENTRY-LEVEL SURVEY
EARTH HISTORY

6. Every place on Earth has a unique story of how its rocks and landforms formed (its geological story). What questions might you ask about a place in order to learn about its geological story?

____

____

____

____

____

____

____

____

____

____

Name ______________________________

INVESTIGATIONS 1–2 I-CHECK
EARTH HISTORY

Date ____________ Class ____________

1. Some students are testing rock samples they collected. If you wanted to test these rocks for calcite, what would you put in the bottle and what would you expect to happen if calcite is there?

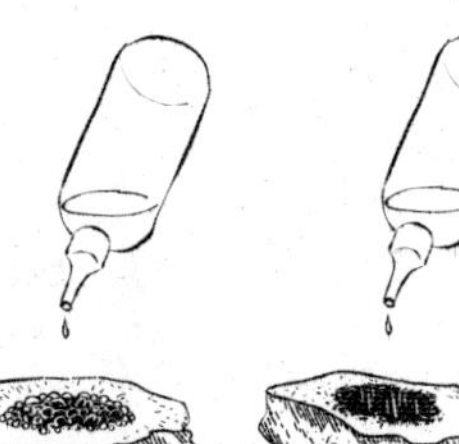

(Mark the one best answer.)

- ❍ **A** Put water in the bottle; if there is calcite, the rock will dissolve.
- ❍ **B** Put salt water in the bottle; if there is calcite, the rock will turn a darker color.
- ❍ **C** Put acid in the bottle; if there is calcite, the rock will fizz.
- ❍ **D** Put liquid soap in the bottle; if there is calcite, the liquid will form a dome on the rock.

2. Consider the following diagram of a river flowing through a landscape. A rain storm causes a large amount of water to flow through the river in a very short time.

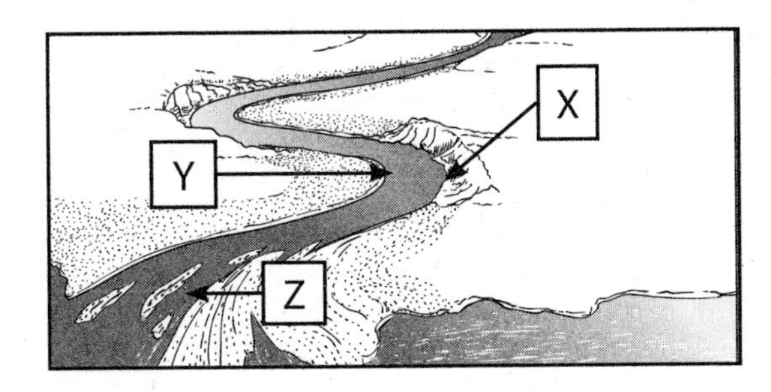

Key
X points to the outside of the curve where the water runs faster.
Y points to the inside of the curve where the water runs slower.
Z points to the mouth of the river where it empties into the ocean.

Which size sediments would you most likely find at each location, X, Y, and Z? Explain why.

____

____

____

____

____

____

____

____

Name ____________________

INVESTIGATIONS 1–2 I-CHECK
EARTH HISTORY

3. The diagram below shows two rock columns.

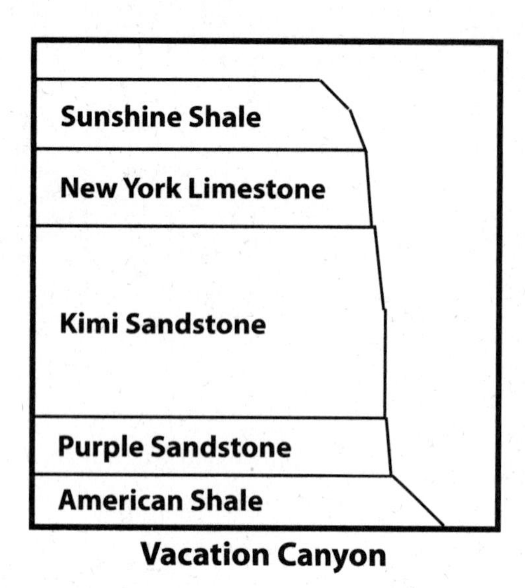

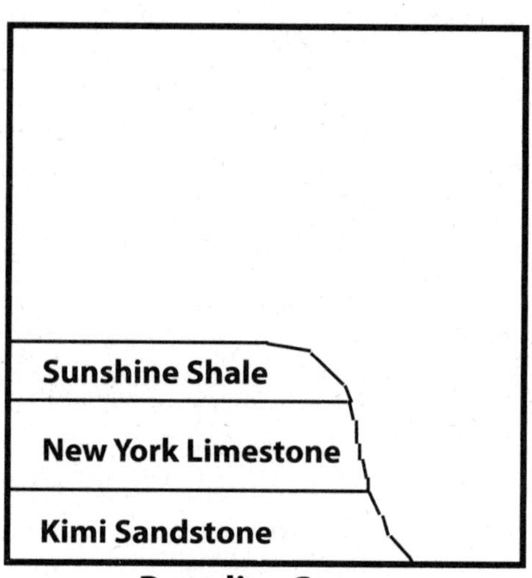

a. A mining company is sending a team to Paradise Canyon to drill down below Kimi Sandstone. What kind of rock do you think they will find directly under the Kimi Sandstone?

(Mark the one best answer.)

❍ **A** Sunshine Shale

❍ **B** New York Limestone

❍ **C** Purple Sandstone

❍ **D** American Shale

b. Surprise Canyon is located between these two canyons, 20 kilometers away from Vacation Canyon and 15 kilometers away from Paradise Canyon. What kind of rock do you think you might find at the top of Surprise Canyon?

(Mark the one best answer.)

❍ **F** Sunshine Shale

❍ **G** New York Limestone

❍ **H** Purple Sandstone

❍ **J** American Shale

Name ______________________________

INVESTIGATIONS 1–2 I-CHECK

EARTH HISTORY

4. Which explanation best describes how a boulder can turn into silt as it moves down a river?

   (Mark the one best answer.)

   ❍ **A** Water washes over the boulder until it is silt-sized.

   ❍ **B** The boulder hits other rocks or objects and breaks apart.

   ❍ **C** The boulder moves down the river until it lands in a basin and becomes silt.

   ❍ **D** A chemical reaction turns the boulder into silt.

5. You find some soil that is mostly made of clay and silt, but no sand. How might this soil have formed?

   (Mark the one best answer.)

   ❍ **A** Slow-moving water carried the sand downstream and left behind the silt and clay.

   ❍ **B** Wind blew all the sand away, leaving behind the silt and clay particles.

   ❍ **C** Silt, clay, and organic material settled out of slow-moving water.

   ❍ **D** A fast-moving stream washed away all the sand.

6. You dig up two soil samples and forget to record where they came from. If you return to the soil samples a week later, what are the characteristics of the soils you would use to match them to the locations where they were collected?

   Write **Y** next to each characteristic that would help identify the soil sample's location; write **N** next to each characteristic that would not be helpful.

   _____ Volume of the sample

   _____ Kinds of sediments

   _____ Amount of moisture

   _____ Color of the sample

   _____ Amount of organic material

Name ______________________

INVESTIGATIONS 1–2 I-CHECK

EARTH HISTORY

7. Use the cross-section drawing to answer the following items.

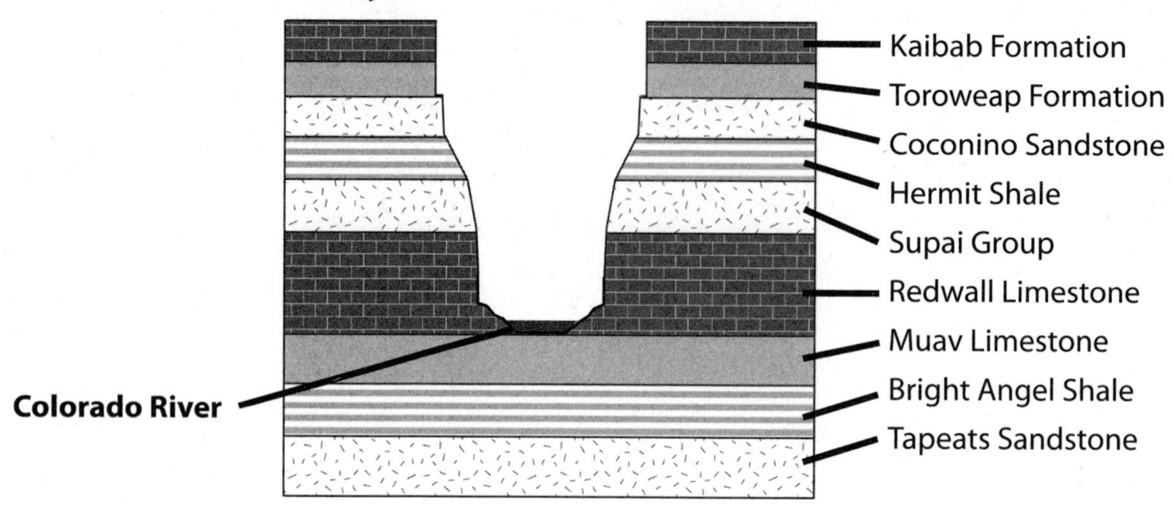

a. The Colorado River started eroding the Grand Canyon 6 million years ago. Of the layers shown at location X, which two layers did the Colorado River erode away first? (*Mark the one best answer.*)

❍ **A** Kaibab Formation and Toroweap Formation

❍ **B** Redwall Limestone and Supai Group

❍ **C** Bright Angel Shale and Tapeats Sandstone

❍ **D** All layers are eroded at the same time by the Colorado River.

b. If you could return to the Grand Canyon in 1 million years, you might be able to see the Mauv Limestone or the Bright Angel Shale at water level. What variables would determine how much the river has eroded in that million years?

____

____

____

____

____

____

Name ____________________

INVESTIGATIONS 3–4 I-CHECK

EARTH HISTORY

Date __________ Class __________

1. Two students are exploring a creek in their neighborhood.

   a. The students hope to find some fossils along the way. What kinds of rocks should the students look for on their hike, and why should they look for those kinds of rock?

   ____

   ____

   ____

   ____

   ____

   ____

   b. In one area, the creek has exposed a layer of shale. What kind of environment might have once existed here to create that rock layer?
   (Mark the one best answer.)

   ❍ **A** A sandy beach

   ❍ **B** A swamp

   ❍ **C** A coral reef

   ❍ **D** A river

   c. Continuing along the creek, they find some pieces of white sandy rock in the creek. They drip hydrochloric acid on that rock. The rock does not fizz.

   Write **Y** next to each process that likely formed the rocks in the creek; write **N** next to each process that did not contribute.

   _____ Deposition

   _____ Chemical reaction

   _____ Cementation

   _____ Erosion

   _____ Melting

Name ______________________________

INVESTIGATIONS 3–4 I-CHECK

EARTH HISTORY

2. a. Which statement best describes the principle of uniformitarianism? *(Mark the one best answer.)*

   ❍ **A** Deposition occurs faster today than it did in ancient environments.

   ❍ **B** Rocks form by the same processes today as in the past.

   ❍ **C** Weathering and erosion can be fast or slow, depending on the type of rock.

   ❍ **D** Earth's geological features were formed by a few catastrophic events.

   b. Why is uniformitarianism an important principle to consider when making inferences about how Earth's geology has changed over millions of years?

   ____

   ____

   ____

   ____

   ____

   ____

   ____

3. What are reasons for gaps in the fossil record?

   Write **Y** next to each statement that explains gaps in the fossil record; write **N** next to each statement that does not explain the gaps.

   _____ Many organisms do not have structures that can be fossilized.

   _____ There were periods of time when no organisms died.

   _____ Some environments do not provide conditions favorable to fossil formation.

   _____ There are lots of fossils that we haven't found yet.

   _____ Some geological events destroyed fossils that were in the rock layers.

Name ______________________________

INVESTIGATIONS 3–4 I-CHECK

EARTH HISTORY

4. Geological time is described in eras and periods. Those segments of time are based on _______.

   (Mark the one best answer.)

   ❍ **A** whoever discovered them

   ❍ **B** significant evolutionary events in Earth's history

   ❍ **C** the thickness of certain rock layers

   ❍ **D** the number of layers of rock that can be counted in 10 meters of a rock column

5. To a geologist, the most important use of index fossils is _______.
   (Mark the one best answer.)

   ❍ **A** to show what kind of life existed long ago

   ❍ **B** to show that the sediments were deposited horizontally

   ❍ **C** to determine the age of the rock

   ❍ **D** to tell the story of the ancient environment

6. Do scientists think that new rocks are still forming today? How does uniformitarianism help scientists answer the first question?

____

____

____

____

____

____

____

____

____

Name ______________________

INVESTIGATIONS 3–4 I-CHECK
EARTH HISTORY

7. Geologists have discovered the fossils you see below in rock layers at two state parks. One of these fossils is an index fossil.

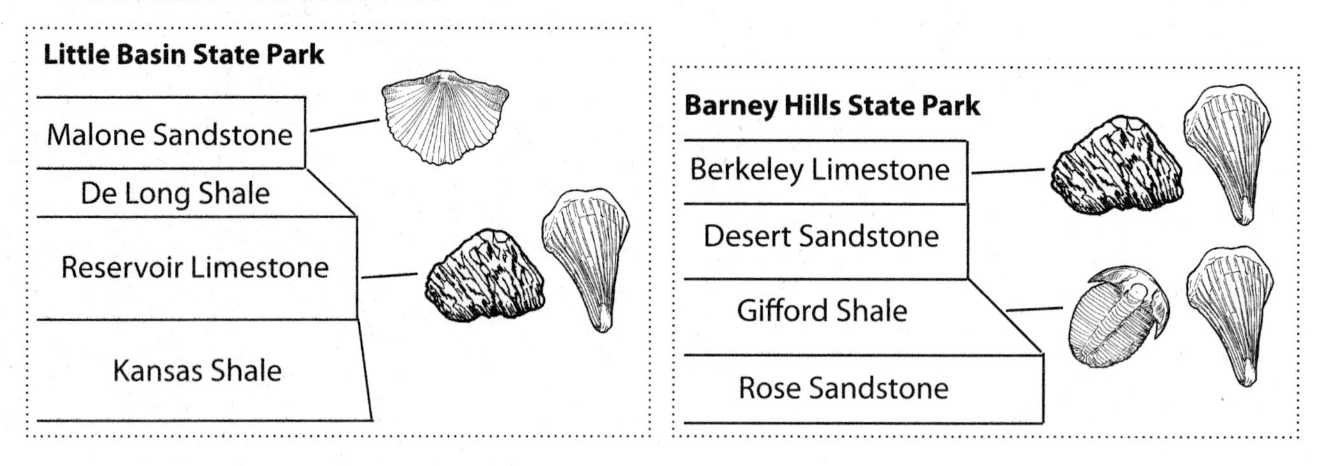

a. Based on the information given, which of the following is most likely the index fossil? *(Mark the one best answer.)*

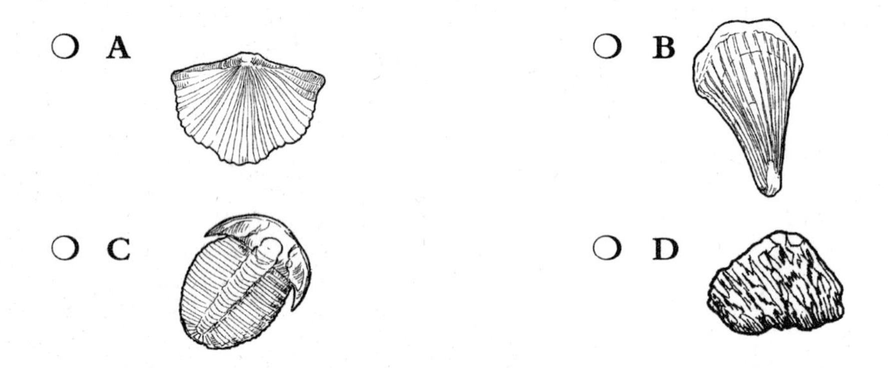

b. Two students used the index fossil to analyze the rock layers. Student X claimed that Reservoir Limestone was older than Desert Sandstone. Student Y disagreed and claimed it was the opposite (Desert Sandstone is older than Reservoir Limestone).

Which student to you agree with, and what is your evidence?

____

____

____

____

____

____

Name ______________________________

INVESTIGATIONS 3–4 I-CHECK

EARTH HISTORY

8. Some students wanted to investigate how five different sediments settled in a basin of water. They started with the basin flat on the table and added new sediments each day for three days.

   On the fourth day, they gently tilted the basin before adding the sediments for the last few days. (Note: The layer shaded gray represents the water.)

   a. Which drawing best represents the basin layers and water at the end of their investigation? *(Mark the one best answer.)*

❍ **A**

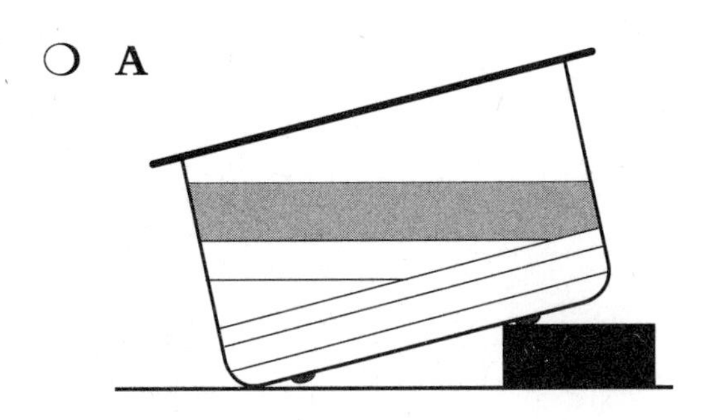

❍ **B**

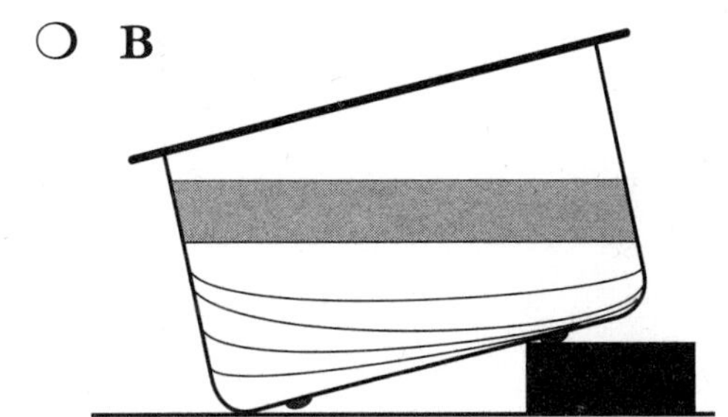

❍ **C**

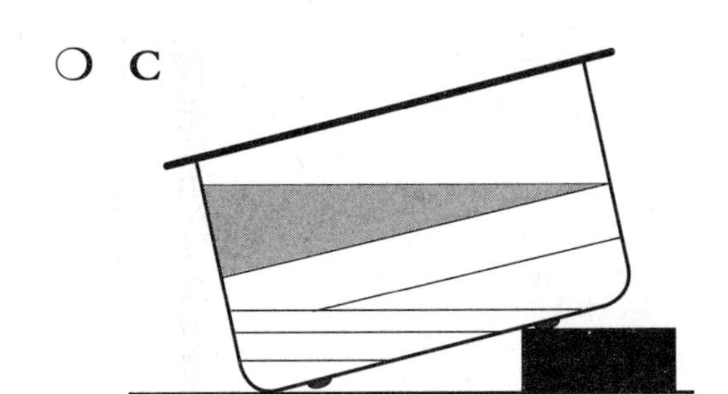

❍ **D**

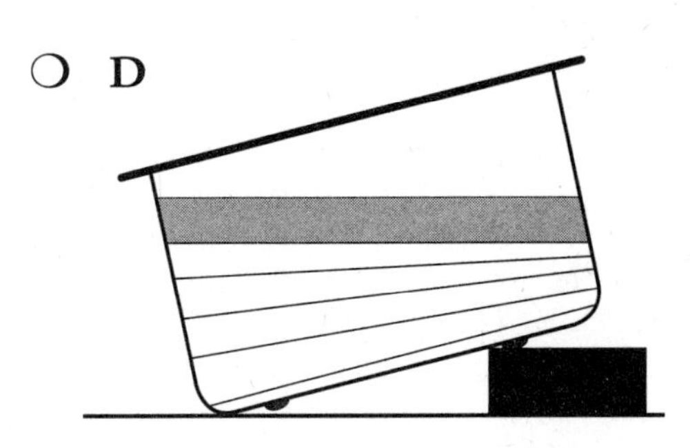

   b. How does this investigation apply to the formation of rock layers in the real world?

   Write **Y** next to each statement that explains how this investigation applies to the formation of rock layers; write **N** next to each statement that does not explain it.

   _____ An entire basin can become tilted.

   _____ Layers always form horizontally.

   _____ Layers gradually tilt until they become horizontal.

   _____ Newer layers form on top of older layers.

   _____ Larger sediments are found in the bottom layers.

Name ___________________________

INVESTIGATIONS 5–6 I-CHECK
EARTH HISTORY

Date __________ Class __________

1. There is an area in Idaho called Craters of the Moon, where most of the ground is covered with basalt, a black igneous rock with no visible crystals. Which statement best describes how the rocks likely formed in this area?

   (Mark the one best answer.)

   ❍ **A** The rocks cooled slowly below ground, and tiny crystals formed.

   ❍ **B** The rocks were under tremendous heat and pressure, so no crystals formed.

   ❍ **C** The rocks cooled quickly above ground forming crystals too small to see.

   ❍ **D** Basalt rocks do not form crystals, no matter how they cooled.

2. How can an understanding of plate tectonics help humans minimize hazards caused by earthquakes and volcanoes?

   ____

   ____

   ____

   ____

   ____

   ____

   ____

   ____

3. A student shared an igneous rock he bought that has beautiful fish and plant fossils in it. Another student gently pointed out that he may want to try to get his money back. How did the student know the rock was fake?

   (Mark the one best answer.)

   ❍ **A** Fish and plants don't fossilize in the same rocks.

   ❍ **B** It is impossible for fossils to form in igneous rocks.

   ❍ **C** It is not legal to sell real fossils.

   ❍ **D** Igneous rocks don't form underwater.

Name ______________________

INVESTIGATIONS 5–6 I-CHECK
EARTH HISTORY

4. The following diagram shows the five major layers of Earth.

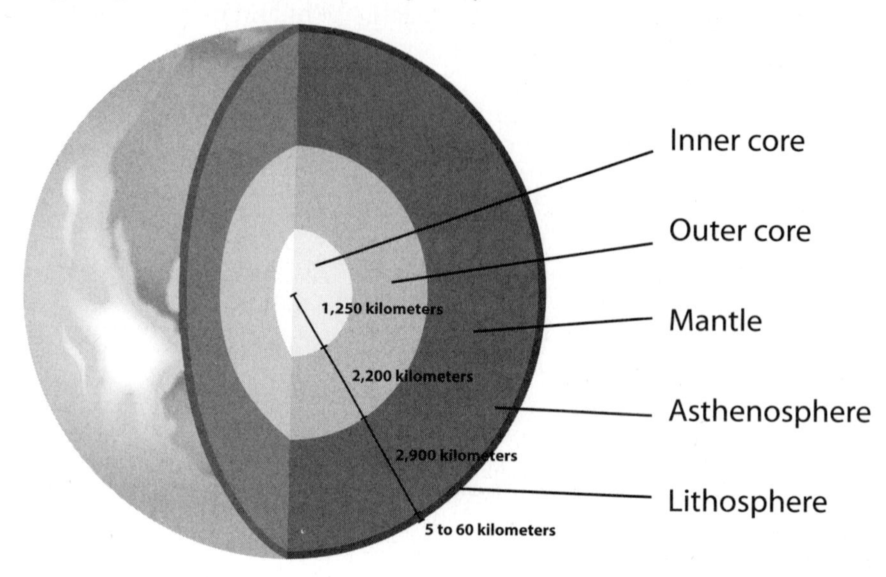

a. Write **A** next to each statement that applies to the asthenosphere; write **L** next to each statement that applies to the lithosphere.

_____ This layer drives the movement of the plates.

_____ The upper mantle is found in this layer.

_____ This layer is semisolid.

_____ Earthquakes happen in this layer.

_____ Convection happens in this layer.

_____ Tectonic plates are in this layer.

b. What is a limitation of this model? Explain how that limitation might affect someone's understanding of Earth's inner structure.

____

____

____

____

____

Name ______________________________

INVESTIGATIONS 5–6 I-CHECK

EARTH HISTORY

5. Read each statement about plate tectonics. Write **T** if the sentence is true; write **F** if the sentence is false.

_____ Volcanoes and earthquakes are often found near plate boundaries.

_____ The continents are made of plates, but not the seafloor.

_____ Scientists can predict where many volcanoes will erupt and earthquakes will occur.

_____ All volcanic eruptions are caused by plate tectonics.

_____ All earthquakes are caused by plate movement.

6. Which question is the best one to ask in order to learn how an igneous rock formed? *(Mark the one best answer.)*

❍ **A** What color is the rock?

❍ **B** What size are the crystals in the rock?

❍ **C** What is the rock made of?

❍ **D** How heavy is the rock?

7. What causes Earth's plates to move?

Write **Y** next to each statement that identifies a cause of plate movement; write **N** next to each statement that is not a cause of plate movement.

_____ The plates are part of the lithosphere.

_____ Different plates have different densities.

_____ Convection currents flow in the asthenosphere.

_____ Subducting oceanic plates pull plates apart.

_____ New crust forms where two plates move away from each other.

Name ____________________

INVESTIGATIONS 5–6 I-CHECK
EARTH HISTORY

8. After hearing about what a student learned in school about plate tectonics, a neighbor responded, "I have a hard time believing that we are still moving around like that."

   What evidence could the student use to support an argument that plate movement is still happening today?

   ____
   ____
   ____
   ____
   ____
   ____
   ____
   ____

9. Consider the following rock samples. Both samples are igneous rocks. Intrusive igneous rocks cool underground, while extrusive igneous rocks cool above ground.

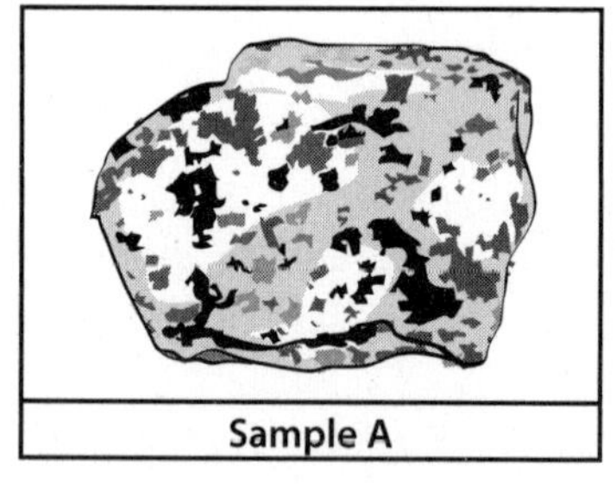

Sample A

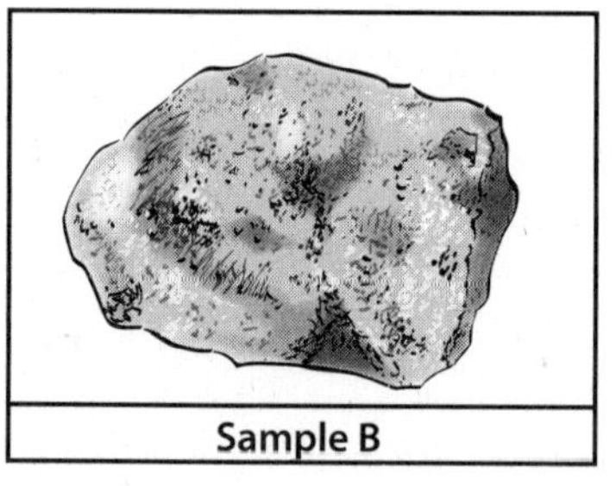

Sample B

   Which sample is an intrusive igneous rock, and which sample is an extrusive igneous rock? What feature of the rocks provides evidence to support your argument?

   ____
   ____
   ____
   ____
   ____

Name ______________________

Date __________ Class __________

1. Mount Moran is a tall mountain in Wyoming. The mountain formed when the North American Plate was uplifted far from the plate boundary. Mount Moran is composed mostly of granite and gneiss. It also has a small patch of sandstone on the top of the mountain and an igneous dike running up through the mountain to the peak.

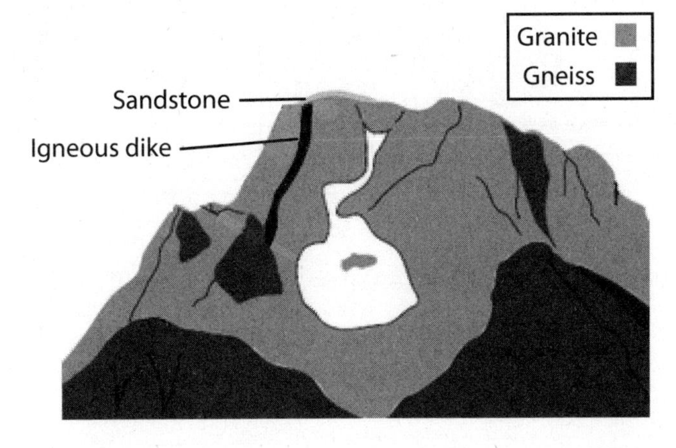

   a. Which is older, the granite or the igneous dike? Why do you think so?

   ____

   ____

   ____

   ____

   ____

   ____

   b. The sandstone found at the peak of the mountain is approximately 3,840 meters above sea level. How can you explain the sandstone on the top of the mountain?

   ____

   ____

   ____

   ____

   ____

   ____

Name ______________________

INVESTIGATION 7 I-CHECK
EARTH HISTORY

2. Consider the following rock samples. Both samples are made of the same minerals.

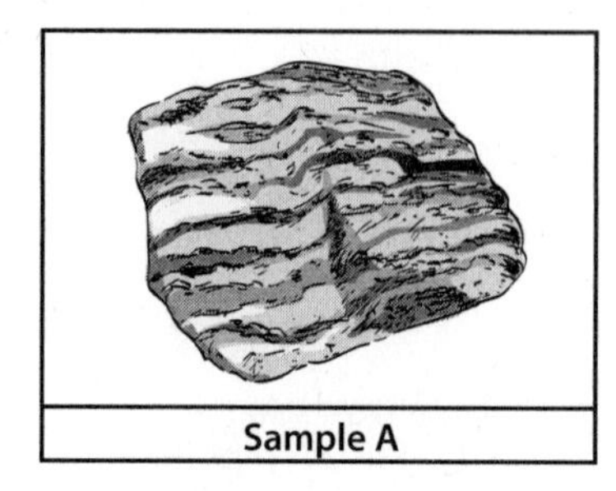
Sample A

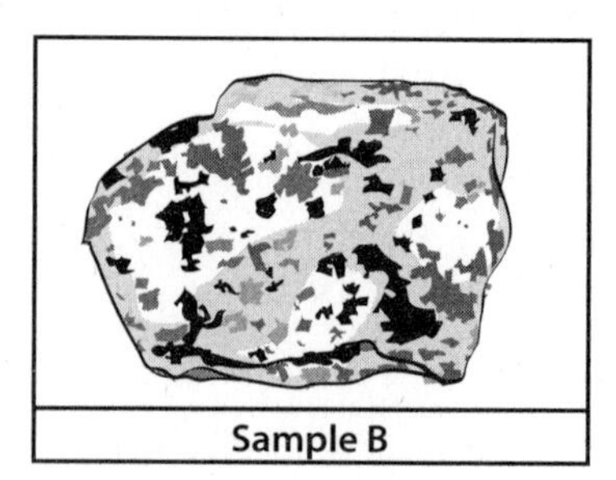
Sample B

a. One sample is an igneous rock, and the other is a metamorphic rock. Which one is which? What feature from the samples provides evidence to support your argument?

____

____

____

____

____

____

b. What processes formed the layers in sample A compared to the processes that formed the layers in sample C?

Sample C: Sedimentary

____

____

____

____

____

____

Name ______________________

INVESTIGATION 7 I-CHECK
EARTH HISTORY

3. The diagram represents a model of two plates of Earth's lithosphere.

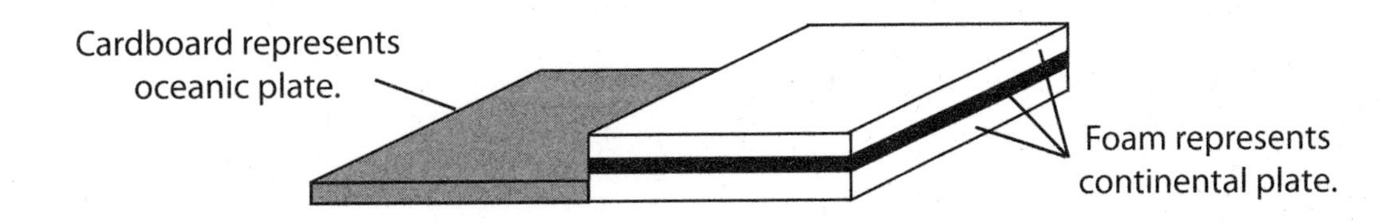

a. What are the possible interactions between these two plates?
Write **Y** next to each statement that could happen when these plates interact; write **N** next to each statement that could not happen.

_____ The plates converge, and both rise high above sea level as they compress.

_____ The continental plate subducts below the oceanic plate.

_____ A rift forms between the two plates as they diverge.

_____ No movement occurs between the two plates.

b. What are the limitations of this model for explaining what happens at plate boundaries?

____

____

____

____

c. Assuming the two plates converge, at which locations would volcanoes likely emerge?

(Mark the one best answer.)

❍ **A** Locations X and Y

❍ **B** Location Y only

❍ **C** Locations X and Z

❍ **D** Location Z only

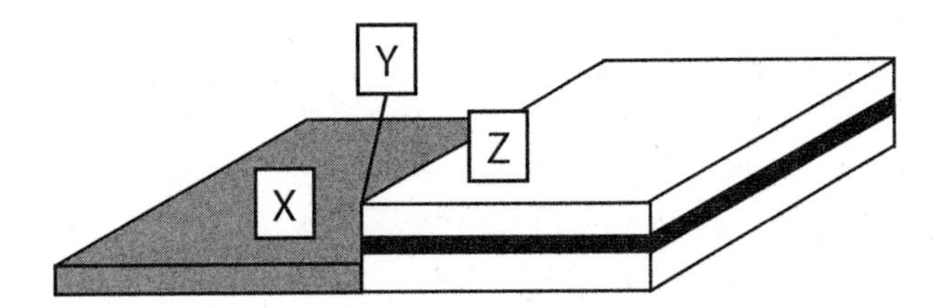

Name ______________________

INVESTIGATION 7 I-CHECK
EARTH HISTORY

4. Scientists recently discovered a patch of oceanic crust deep beneath the Mediterranean Sea that is estimated to be 340 million years old. It is the oldest oceanic crust ever found.

   a. If scientists believe Earth is actually 4.6 billion years old, why haven't scientists found older oceanic crust?
   (Mark the one best answer.)

   ❍ **A** Rocks did not cool on Earth until about 340 million years ago.

   ❍ **B** It is too hard to collect samples of oceanic crust.

   ❍ **C** Oceanic crust subducts under continental crust as new oceanic crust forms.

   ❍ **D** Oceanic crust is younger than continental crust.

   b. The oldest surface Earth rocks that scientists have found are continental crust and estimated to be 3.8 billion years old.

   Why have scientists been able to find rocks in continental crust so much older than the oldest rocks found in oceanic crust?

   (Mark the one best answer.)

   ❍ **F** It is much harder to collect samples of oceanic crust.

   ❍ **G** Less-dense continental crust is not subducted.

   ❍ **H** Continental crust cooled and formed long before oceanic crust.

   ❍ **J** Oceanic crust is harder to date than continental crust.

5. For a sedimentary rock to change into another kind of sedimentary rock, it would first need to _____.

   (Mark the one best answer.)

   ❍ **A** melt

   ❍ **B** compress under pressure

   ❍ **C** be weathered

   ❍ **D** be deposited somewhere

Name ______________________________

POSTTEST
EARTH HISTORY

Date ___________ Class ___________

1. Consider the map below.

   a. On the map, place at least five more X marks to indicate places you are most likely to find volcanoes.

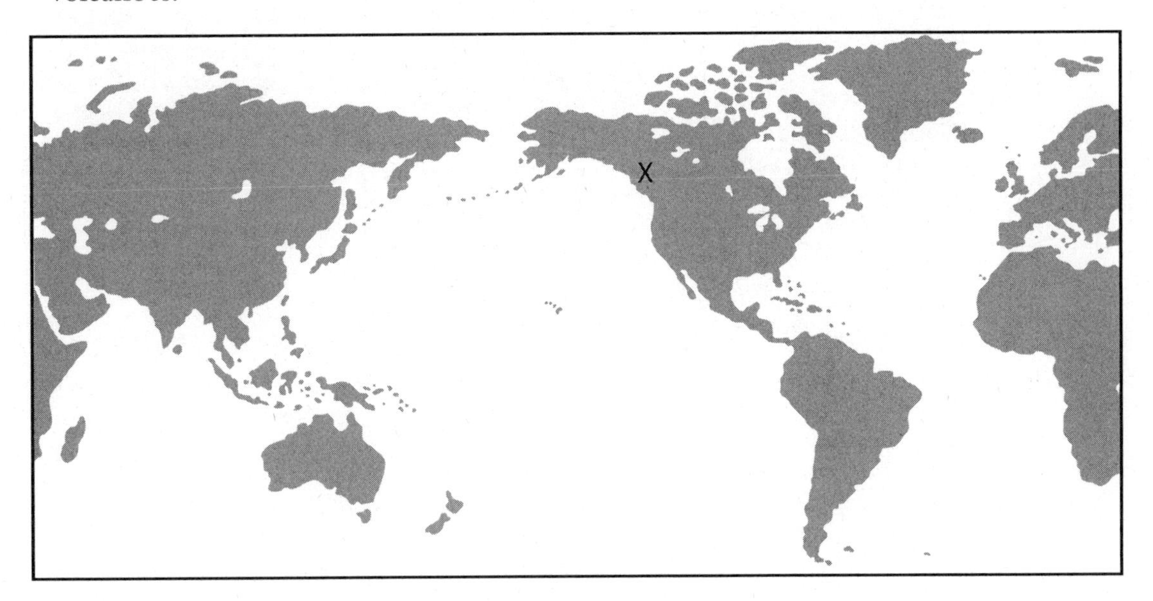

   b. Where are you most likely to experience an earthquake?

   (Mark the one best answer.)

   ❍ **A** Along a volcanic island chain

   ❍ **B** Along the boundary of a crustal plate

   ❍ **C** In a subduction zone

   ❍ **D** Randomly anywhere on Earth

2. What causes plates to move?
   (Mark the one best answer.)

   ❍ **A** Earthquakes and volcanoes

   ❍ **B** Thermal energy from the core

   ❍ **C** Sinking of the lithosphere into the asthenosphere

   ❍ **D** Plate interactions

Name ______________________

POSTTEST
EARTH HISTORY

3. Consider these three samples of different types of rocks.

Sample A: Igneous (Granite)	**Sample B: Metamorphic (Gneiss)**	**Sample C: Sedimentary (Sandstone)**

a. In the space below, draw a diagram to explain how earth materials can cycle among these three types of rocks.

Sample D: Igneous (Obsidian)

b. Consider sample D, another igneous rock with a glassy surface. Which cooled faster, sample A or sample D? What feature provides evidence to support your argument about which one cooled faster when it formed?

____

____

____

____

____

4. Look at the river in the illustration. The water at F is moving much faster than at G. The bank is much deeper on the outside of the river bend (F) than it is on the inside of the river bend (G).

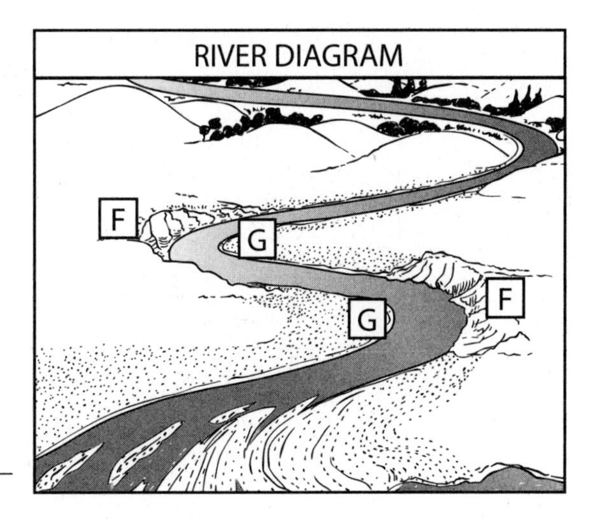

Compare the two sides of the river.

a. How does the faster current at F affect erosion and deposition at the riverbank?

____

____

____

____

b. How does the slower current at G affect erosion and deposition at the riverbank?

____

____

____

c. Assume that for the next 1,000 years, the only geological processes affecting this area are erosion and deposition from the river. What might you expect to see at the end of 1,000 years?

(Mark the one best answer.)

❍ **A** The river will be straighter.

❍ **B** The river will have more dramatic curves.

❍ **C** The delta will have dammed the river.

❍ **D** The river will have cut a deep canyon.

Name ____________________

POSTTEST
EARTH HISTORY

5. Use the illustration of a rock column to answer the following questions.

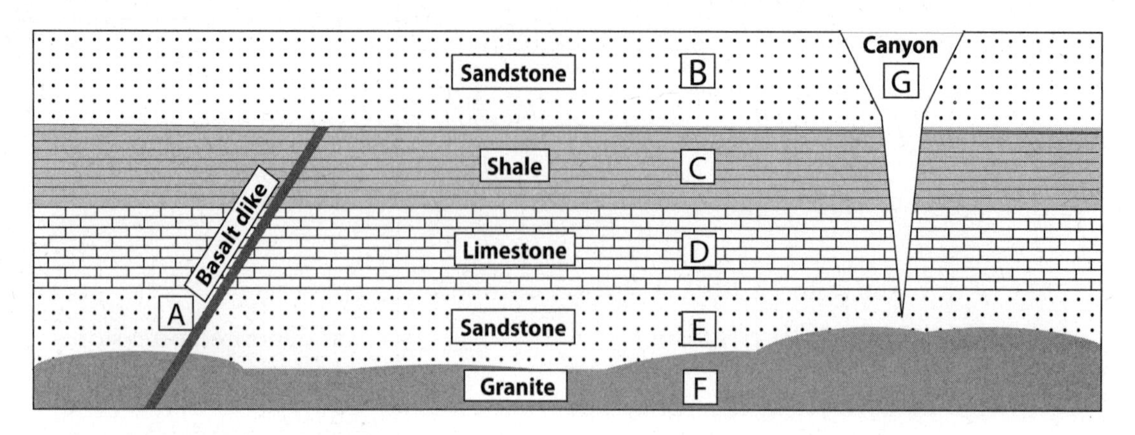

a. Which feature is youngest?
(Mark the one best answer.)

- ❍ **A** Dike (A)
- ❍ **B** Canyon (G)
- ❍ **C** Shale (C)
- ❍ **D** Limestone (D)

b. Which feature is oldest?
(Mark the one best answer.)

- ❍ **F** Dike (A)
- ❍ **G** Canyon (G)
- ❍ **H** Shale (C)
- ❍ **J** Limestone (D)

c. Using various dating techniques, geologists have calculated the following ages for two rocks in the rock column.

Rock	Age
Rock A Basalt	3 million years old
Rock E Sandstone	200 million years old

Based on the data in the table, how old is rock B?
(Mark the one best answer.)

- ❍ **A** Between 200,000 to 3 million years old
- ❍ **B** Less than 3 million years old
- ❍ **C** Between 3 million to 200 million years old
- ❍ **D** More than 200 million years old

Name ______________________________

POSTTEST
EARTH HISTORY

6. A geologist finds a rock layer that is tilted. The fossils in the rock include oyster shells and brachiopods. It fizzes when acid is dropped on it. Evidence supports that this rock ________.

   Write **Y** next to each statement that can be supported by the evidence; write **N** next to each statement that cannot be supported.

   _____ is a sedimentary rock

   _____ contains salt deposits

   _____ contains calcite

   _____ was originally part of a sand dune

   _____ was originally horizontal

   _____ was deposited in a marine environment

   _____ was deposited on a slope

7. Scientists have discovered that Mount Everest is growing taller by about 2 centimeters per year.

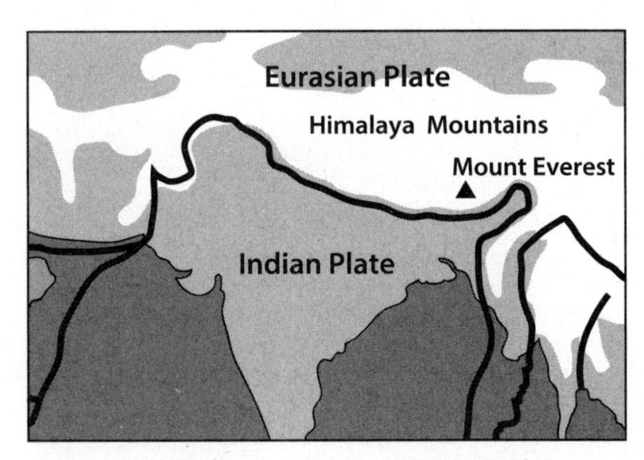

Solid black lines are plate boundaries.

What is a possible scientific explanation for this?

(Mark the one best answer.)

❍ **A** The Indian Plate and Eurasian Plate are converging and pushing the Himalaya mountains upward.

❍ **B** The Indian Plate is subducting under the Eurasian Plate, pushing the Himalaya mountains upward.

❍ **C** The Eurasian Plate is subducting under the Indian Plate, pushing the Himalaya mountains up.

❍ **D** The Eurasion and Indian Plates are moving past each other, pushing the Himalaya mountains up.

Name ____________________________

POSTTEST
EARTH HISTORY

8. Fault lines or cracks can be found throughout the Rocky Mountains, and all three types of rocks are found there. What does this tell you about these mountains?

   Write **Y** next to each statement that is supported by the evidence; write **N** next to each statement that is not supported.

   _____ An ocean covered this part of the country before the mountains formed.

   _____ A great ice sheet once covered this part of the country.

   _____ This part of the North American Plate has experienced intense pressure.

   _____ Faults allowed melted rock to come to the surface.

9. Mountains of the British Isles and the Caledonian Mountains of Scandinavia have the same rock composition as the Appalachian Mountains.

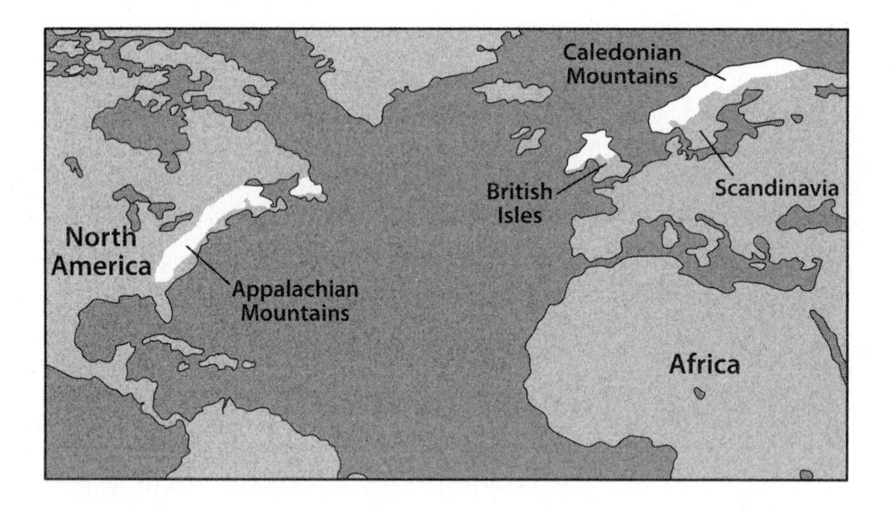

   What does this tell us about the story of the past geology of this part of Earth?

   ____

   ____

   ____

   ____

   ____

Name ______________________

POSTTEST
EARTH HISTORY

10. James Hutton was a Scottish geologist and experimental farmer. He would often think about how the land on his farm changed over time.

    a. Which processes formed the soil on his land?

    Write **Y** next to each process that likely helped form the soil; write **N** next to each process that did not help.

    _____ Weathering

    _____ Melting

    _____ Deposition

    _____ Crystallization

    _____ Decomposition

    b. Many scientists of Hutton's day thought Earth's landforms had formed during one sudden event, then didn't change. Hutton watched how the stream running through the hills on his farm carried soil away after each rainstorm. "If soil is always being carried away," he wondered, "why was the land not completely flat by now?"

    If you were a modern-day scientist who was able to go back in time to have a conversation with Hutton at the moment he made this observation, what would you tell him?

    ____

    ____

    ____

    ____

    ____

    ____

    ____

    ____

    ____

Notebook Answers

Landforms Tour

	Site name	Landforms, observations, and questions
1	My school	Answers should include
2	Mount St. Helens	volcano, caldera, mountains, lake
3	Aleutian Islands	island, mountains, coastline, ocean
4	Alaskan glaciers	glacier, river, mountains
5	Na Pali Coast	coastline, mountains, ocean, island
6	Death Valley	mountains, rivers, fans
7	Florida Keys	islands, reef, ocean
8	Stone Mountain	river, mountain, lake
9	Shenandoah	mountains, valleys
10	Central Park	lakes, rivers, coastline
11	Niagara Falls	waterfalls, river, cliffs
12	Delta	coastline, river, lake, delta
13	Mississippi River	river, floodplains, delta, meanders
14	Lake Ozark	lake, river
15	Oklahoma Panhandle	plains, floodplain, river
16	Missouri River	plains, floodplain, river, meanders
17	Grand Tetons	mountains, plains, river, valleys
18	Grand Canyon	canyon, rivers, cliffs, plateau

Anticipation Guide

Before reading the article called "Seeing Earth," check off the statements with which you agree.

____ 1. Images in Google Earth™ show live images of Earth as it appears right now.

____ 2. Rocks are actual pieces of Earth.

____ 3. Google Earth™ shows representations of the surface of planet Earth.

____ 4. Earth-imaging satellites can take a picture of the whole surface of one side of Earth.

After reading the article, check off the statements that were confirmed in the text.

____ 1. Images in Google Earth™ show live images of Earth as it appears right now.

__X__ 2. Rocks are actual pieces of Earth.

__X__ 3. Google Earth™ shows representations of the surface of planet Earth.

____ 4. Earth-imaging satellites can take a picture of the whole surface of one side of Earth.

Write a few sentences explaining how this article helped you clarify your thinking about Earth imaging.

WARNING — This set contains chemicals that may be harmful if misused. Read cautions on individual containers carefully. Not to be used by children except under adult supervision.

Mile 20 Rock Observations

MILE 20 ROCK OBSERVATIONS

Rock number	Color	Texture	Observations from photos	Other

Answers will vary.

Mile 20 Sketch

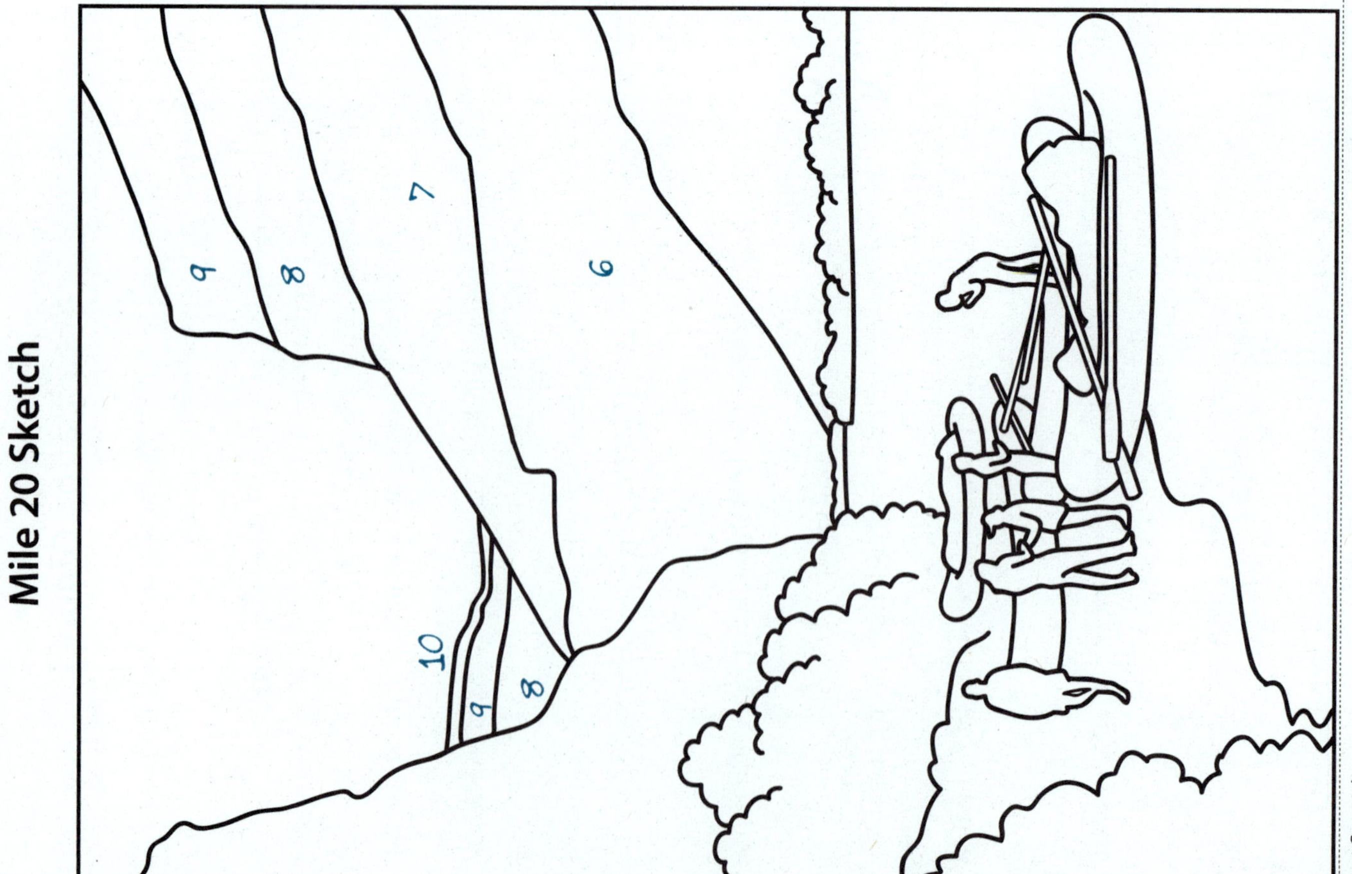

Mile 52 Sketch

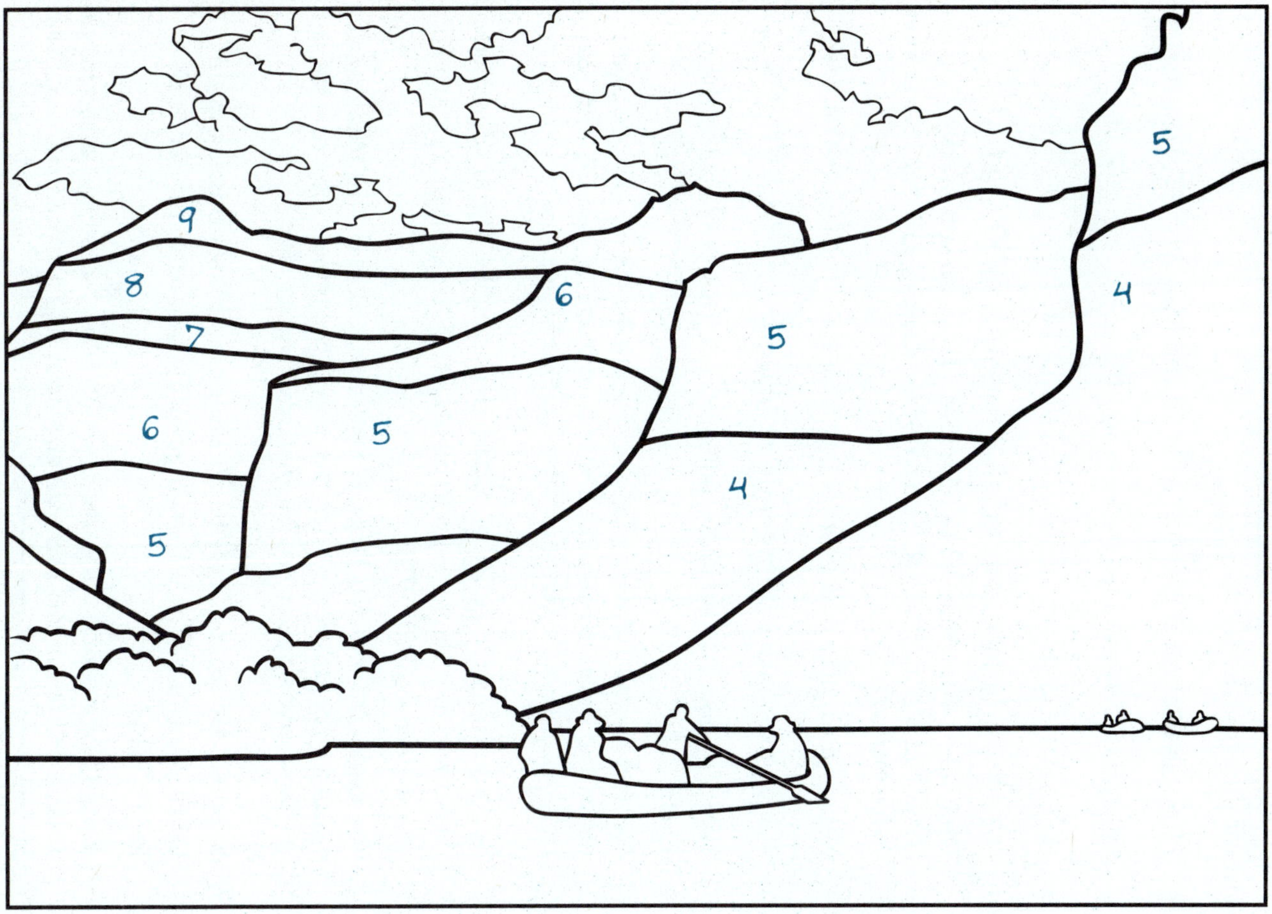

Mile 52 Rock Observations

MILE 52 ROCK OBSERVATIONS

Rock number	Color	Texture	Observations from photos	Other

Answers will vary.

Grand Canyon Rocks

Mile 52	
Rock	**Rock-layer name**
9	Limestone
8	Sand
7	Shal
6	Sandstone
5	Limestone
4	Limestone
Colorado River	
Elevation of river: 853 meters	

Mile 20	
Rock	**Rock-layer name**
	Limestone
	Limestone
8	Sandstone
7	Shale
6	Sandstone
Colorado River	
Elevation of river: 892 meters	

See Investigation 1, Part 3, for an example of a finished sheet.

Correlation Questions

Use your "Grand Canyon Rock Lineup" and *FOSS Science Resources* to answer these questions.

1. How far apart are the sites at Mile 20 and Mile 52? 32 miles (51 km)
2. What is the elevation of the river at Mile 20? 892 m
3. What is the elevation of the river at Mile 52? 853 m
4. Which way is the Colorado River flowing, from Mile 20 to Mile 52 or from Mile 52 to Mile 20? How do you know?
   From Mile 20 to Mile 52. Water flows downhill, and Mile 52 is lower than Mile 20.
5. Which rock layer is at river level at Mile 20? Supai Sandstone
6. Which rock layer is at river level at Mile 52? Muav Limestone
7. Why do these two sites have different rock layers exposed at river level?
   The rock layers are horizontal. The Muav Limestone is a rock layer under the Supai Sandstone. The river is lower at Mile 52 and has cut through more layers and into the Muav Limestone.
8. Suppose you were in a boat on the river at Mile 20 and you could drill down into the rock under the river. What kind of rock would you expect to find? Why?
   You would drill into the Redwall Limestone and then the Muav Limestone because they are both under the Supai Sandstone. You can see both of these layers under the Supai Sandstone at Mile 52.
9. Suppose you stopped at Mile 36 along the Colorado River in the Grand Canyon. Which rock layer would you expect to see at river level? Why?
   Redwall Limestone. It is the layer between the Supai Sandstone and Muav Limestone. The river has cut down into the Redwall Limestone but not into the Muav Limestone.

Stream-Table Map

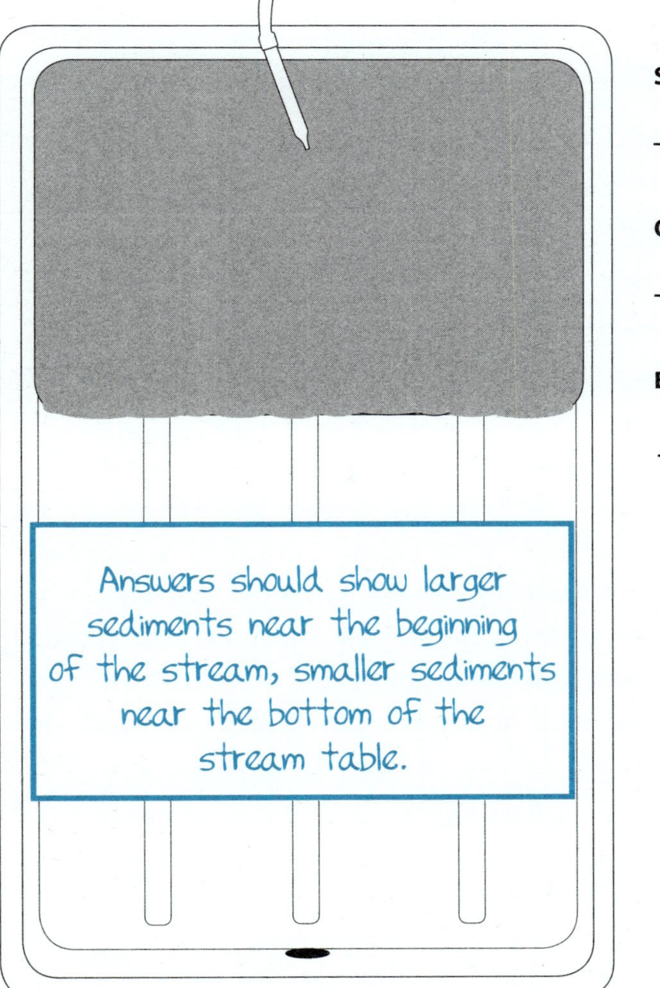

Starting time ____________

Observation time ____________

Elapsed time ____________

1. Observe where different materials are being deposited (that is, sand, clay), and label them on the map.
2. Label on the map where the water is flowing fastest and where it is flowing slowest.
3. Label on the map any landforms that have been created.

Stream-Table Questions

Refer to your *Stream-Table Map* as you answer these questions.

1. Watch a grain of sand as it moves along. Describe its motion.
   It bounces along with the flow of the water until it reaches the flatter part of the tray. Then the sand stops moving.
2. Where are the large particles deposited? The small particles?
   The large particles are deposited near where the water flows off the sand cliff. The small particles are deposited on the flat part of the tray, and some go all the way into the catch basin.
3. Is a delta forming? Where? Why is it forming there?
   Yes, a delta is forming where the water flows off the sand cliff. That is where the big pieces of sand are deposited. The water is more spread out and not moving as fast there.
4. What color is the water flowing into the basin?
   The water flowing into the basin is milky white and very cloudy.
5. Consider the Grand Canyon, and refer to the Colorado Plateau map in *FOSS Science Resources*. Where is the material eroded by the Colorado River deposited today?
   Along the channel in sandbars, along riverbanks, and behind dams. Some might make it all the way to Lake Mead, and get stuck behind the Hoover Dam.
6. Where do you think the material that was eroded by the Colorado River was deposited in the past?
   The eroded material would be farther down the Colorado River in the southwest corner of Arizona. It would have eventually ended up at the mouth of the river in the Gulf of California.
7. Which do you think came first, the Colorado Plateau, the Colorado River, or the Grand Canyon? Support your answer with evidence.
   The Colorado Plateau had to come first. There had to be rocks for the Colorado River to erode. As the river eroded the rocks, the Grand Canyon formed.

Stream-Table Videos

Part 1: Flood

1. Describe in detail the motion of earth materials you see in a flood condition.

   Materials are moving faster down the stream. Larger materials than what we saw in class are being carried by the water.

Part 2: Different earth materials

Homogeneous = mixture of sand and clay just like the material used in class

Heterogeneous = mixture of sand and clay on top and bottom; layer of red clay in the middle

2. In which stream table is the earth material eroding faster and deeper?

   Homogeneous

3. What is happening to the top sand/clay level in the heterogeneous materials?

   The materials are being eroded away by the water.

4. What is happening to the layer of red clay?

   The clay seems to be more sturdy. The water doesn't erode it as quickly as the other materials.

5. The bottom layer was made out of the same material as the top layer. Why didn't it erode as quickly?

   It was protected underneath the layer of the clay, so water couldn't erode it as quickly.

Response Sheet—Investigation 2

A group of students were collecting soil samples near a river. One student noted there was a lot of sand in one of the samples.

He said, "All this sand must have washed up from the beach where the river enters the ocean."

Another student said, "I think it came from the rocks in the mountains. But there aren't any mountains near us. I don't understand how the sand got there."

What would you tell the students to help them solve where the sand came from?

The second student is right: sand comes from rocks that have been broken down. Some of these rocks might come from mountains, but they can also come from other places where rocks have been deposited. Rocks can break apart due to weathering, which is when wind, water, or other forces break rocks into smaller pieces, including sand. These rock pieces can move from mountains and other locations by erosion, which is when wind, water, or other forces move rock pieces. As they get washed or blown away, they settle out in areas like beaches.

Seawater Investigation

WARNING — This set contains chemicals that may be harmful if misused. Read cautions on individual containers carefully. Not to be used by children except under adult supervision.

Materials

1 Plastic cup with lid
60 mL Limewater (calcium hydroxide solution)
4 Straws with hole punched in side
4 Safety goggles

Instructions

1. Work with your group. Measure 60 mL of limewater into a cup. Limewater is a $Ca(OH)_2$ solution.
2. Place the lid on the cup.
3. Use the table below to record observations before bubbling.
4. Take turns poking your straw through the hole in the lid and gently blowing air into the limewater. Continue taking turns for 2 or 3 minutes. **SAFETY NOTE: Don't suck up the limewater.** Make sure you don't blow so hard that the water splatters. If you get some limewater on your hands, rinse them with clear water.
5. Record observations after bubbling.
6. Let the cup stand for 5 minutes and then record your observations.

Seawater Observations

Observations of $Ca(OH)_2$ cup *before* bubbling	Observations of $Ca(OH)_2$ cup *after* bubbling	Observations of $Ca(OH)_2$ cup *after* standing for 5 minutes
The water is clear (or slightly cloudy).	The limewater is very cloudy and milky white. It is difficult to see through it.	Most of the white cloudy material has settled out and formed a white layer on the bottom of the cup.

Analysis: What do you think the limewater reaction has to do with limestone formation?
The precipitate that formed in the cup is like sediment that forms in the ocean and deposits on the bottom, which can eventually become limestone.

Basin Questions

Use the correlation of Grand Canyon rocks from Investigation 1 to help answer these questions.

1. Which Grand Canyon rock layer is the oldest that we have observed so far?
   Muav Limestone
2. How do you know it is the oldest?
   It is the lowest layer of the rocks we have studied. It must have formed first, before everything else was deposited on top.
3. Which layer in the Grand Canyon is the youngest that we have observed so far?
   Kaibab Limestone or Formation
4. How do you know it is the youngest?
   It is the layer found at the top of the Grand Canyon.
5. What do you think is below the oldest layer?
   More rock, maybe in layers.

Response Sheet—Investigation 4

A student's little brother said to her,

"The fossils in the Grand Canyon are dead bodies of animals that washed up onto the rocks as the river passed by."

If you were the student, what would you say to your brother?

The animal bodies weren't washed up by the river, they were in place before the river was there. The rock layers formed with bodies of dead animals in them. Those bodies became fossils that are part of the rock. Eventually, the rock layers were lifted to their current height. The fossils are revealed when the river weathers away the rock.

Geologic Time Calculations

Era	Period	Age (years)	Distance on time line (mm)	Distance on time line (cm)
Cenozoic	(Today) Quaternary	0.00	0.0	0.0
	Tertiary	2,600,000	2.6	0.26
Mesozoic	Cretaceous	66,000,000	66.0	6.6
	Jurassic	145,000,000	145.0	14.5
	Triassic	201,000,000	201.0	20.1
Paleozoic	Permian	252,000,000	255.0	25.2
	Pennsylvanian	299,000,000	299.0	29.9
	Mississippian	323,000,000	323.0	32.3
	Devonian	359,000,000	359.0	35.9
	Silurian	419,000,000	419.0	41.9
	Ordovician	444,000,000	444.0	44.4
	Cambrian	485,000,000	485.0	48.5
Precambrian		541,000,000	541.0	54.1
		4,600,000,000	4,600.0	460.0

1 mm = 1,000,000 years

Index-Fossil Correlations

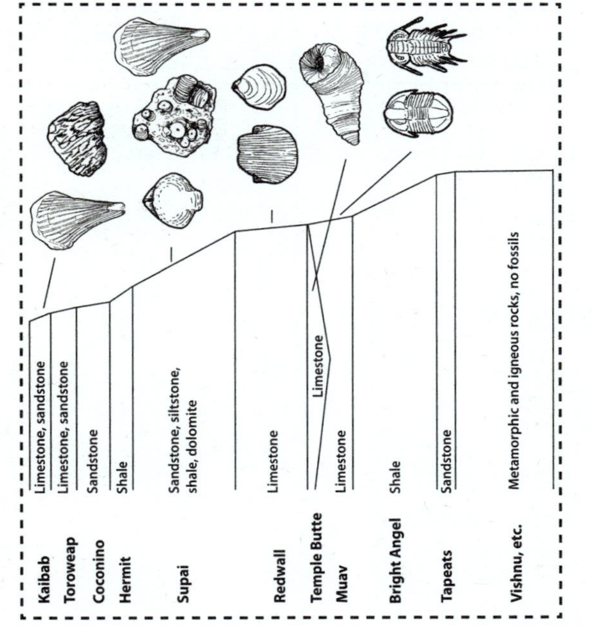

B = Bryce Canyon
Z = Zion National Park

See Investigation 4, Part 3, for an example of a finished, correlated sheet.

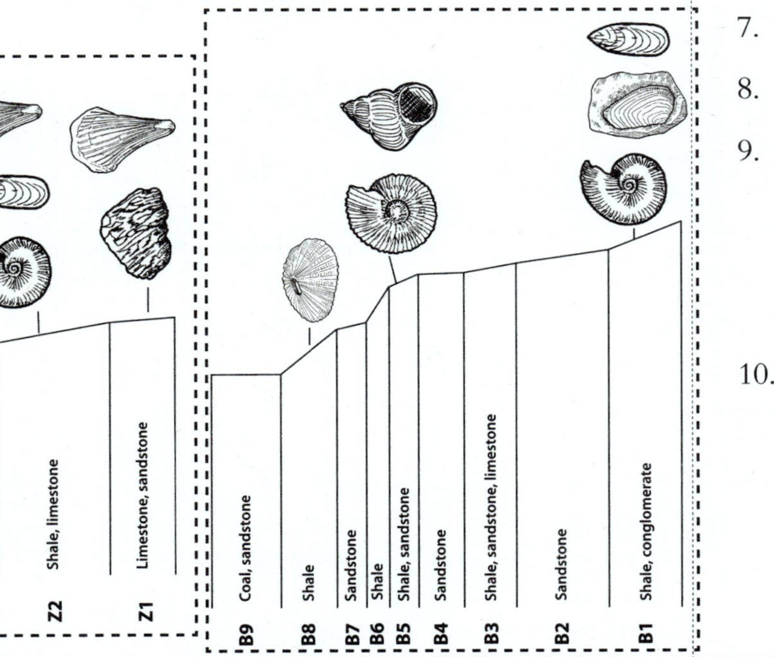

Index-Fossil Correlation Questions

Answer these questions after you have identified and correlated the rock layers at the three parks.

1. Were any layers at all three canyons the same age? No
2. Which rock layers contained the same index fossils at Zion and the Grand Canyon? Kaibab Limestone (Formation) and Z1 (limestone, sandstone)
3. Which rock layers contained the same index fossils at Zion and Bryce? Z7 and B5; Z2 and B1
4. Which rock layers contained the same index fossils at the Grand Canyon and Bryce? None
5. Which canyon has the oldest rocks? The Grand Canyon
6. What was the age of the oldest rock layer? Older than middle Cambrian
7. Which canyon has the youngest rocks? Bryce Canyon
8. What was the age of the youngest rock layer? Younger than late Jurassic
9. Is rock layer B3 at Bryce older or younger than the Supai Group at the Grand Canyon? How do you know? Younger. B3 comes between layers B5 and B1, which contain identifiable index fossils. B5 contains fossils from the late Jurassic. B1 contains fossils from the early Triassic. The Supai Group contains index fossils from the late Pennsylvanian. So B3 has to be older than early Triassic, making it younger than the Supai Group.
10. Is rock layer B2 at Bryce older or younger than rock layer Z1 at Zion? How do you know? Younger. Z1 contains Permian index fossils. B1 contains fossils from the early Triassic, making it younger than Z1. B2 is on top of B1, so it is younger than B1. So it is also younger than Z1.

Rocks over Time

Rock Layer	Time of deposition (approximately)	Distance on time line	Period
Kaibab Formation	Ended 255,000,000 years ago Began 260,000,000 years ago	25.5 cm 26.0 cm	Permian and Triassic
Toroweap Formation	Ended 260,000,000 years ago Began 265,000,000 years ago	26.0 cm 26.5 cm	Permian
Coconino Sandstone	Ended 265,000,000 years ago Began 270,000,000 years ago	26.5 cm 27.0 cm	Permian
Hermit Shale	Ended 270,000,000 years ago Began 275,000,000 years ago	27.0 cm 27.5 cm	Permian
Supai Group	Ended 275,000,000 years ago Began 325,000,000 years ago	27.5 cm 32.5 cm	Pennsylvanian
Redwall Limestone	Ended 325,000,000 years ago Began 360,000,000 years ago	32.5 cm 36.0 cm	Mississippian
Temple Butte Limestone	Ended 370,000,000 years ago Began 375,000,000 years ago	37.0 cm 37.5 cm	Devonian
Muav Limestone	Ended 525,000,000 years ago Began 530,000,000 years ago	52.5 cm 53.0 cm	Cambrian
Bright Angel Shale	Ended 530,000,000 years ago Began 540,000,000 years ago	53.0 cm 54.0 cm	Cambrian
Tapeats Sandstone	Ended 540,000,000 years ago Began 545,000,000 years ago	54.0 cm 54.5 cm	Cambrian

1 cm = 10,000,000 years

Cooling-Rate Investigation

WARNING — This set contains chemicals that may be harmful if misused. Read cautions on individual containers carefully. Not to be used by children except under adult supervision.

What effect does cooling rate have on crystal formation?

1. What variables do you need to consider in the design of your experiment?
2. What materials will you need for your investigation?
3. Describe your procedure.
4. Record data and/or describe your results.
5. What conclusions can you draw about crystals in igneous rocks?

Answers will vary.

Igneous-Rock Observations

WARNING — This set contains chemicals that may be harmful if misused. Read cautions on individual containers carefully. Not to be used by children except under adult supervision.

Rock #	Description	Intrusive or extrusive?	Identification
11	Pink, gray, and/or white, large crystals	Intrusive	Granite
12	Black, no visible crystals	Extrusive	Basalt
16	Black or gray, glassy	Extrusive	Obsidian
17	Gray, lot of holes, glassy, very lightweight	Extrusive	Pumice
19	Black, lot of holes, no visible crystals	Extrusive	Scoria
21	Gray, light colored, grainy and very tiny or no crystals	Extrusive	Tuff
23	Black, large crystals	Intrusive	Gabbro
24	Gray, no visible crystals	Extrusive	Rhyolite

Rock-Layer Age Puzzle

This illustration shows a rock column. Using potassium-argon dating, geologists have calculated an age of 200 million years for rock A, a granite. Rock F, the volcano, has been given an age of 225,000 years. Use the ages and illustration to answer the questions.

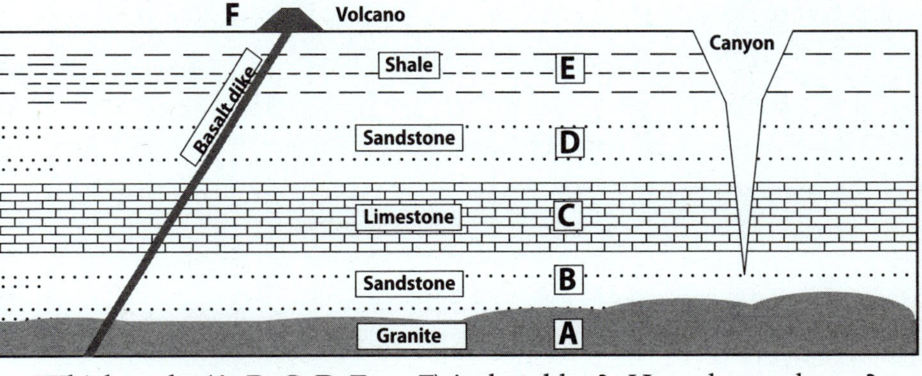

1. Which rock (A, B, C, D, E, or F) is the oldest? How do you know?
   A is the oldest because it is at the bottom, and other layers formed by sedimentation (B–E) or volcanism (F) after it was already there.

2. Which rock (A, B, C, D, E, or F) is the youngest? How do you know?
   F is the youngest. It cuts through all the other layers, so they must have been there first.

3. Did the canyon form before or after layers B, C, D, and E? How do you know? It must have been formed after these layers. The canyon eroded into the rock layers, so the layers had to have been there first.

4. Did rock B form before or after rock C? How do you know?
   B formed before C, because C was deposited on top of B.

5. When did rocks B, C, D, and E form? Give a range, between XXX and XXX. How do you know? Between 200 mya and 225,000 years ago. They must have formed after 200 mya because rock layer A wasn't formed until then, and they formed on top of A. They formed before 225,000 years ago, when F cut through the layers.

6. Which rock layers are sedimentary rocks? Which are igneous rocks?
   B, C, D, and E are sedimentary. The granite in A is igneous. F is igneous.

Earth's Layers Information

Layer	Consistency	Composition	Temperature	Density	Thickness
Crust (oceanic)	Rigid, broken into plates	Basalt	From ocean temperature to 870°C near the mantle boundary	More dense than the continental crust	5 km to 8 km
Crust (continental)	Rigid, broken into plates	Igneous rocks like granite, sedimentary rocks, other rocks	From air temperature to 870°C near the mantle boundary	Less dense than the oceanic crust	30 to 60 km
Mantle (solid upper mantle)	Rigid, broken into plates	More iron and magnesium than the crust; less silicon and aluminum	500°C to 900°C	More dense than crust and asthenosphere	0km to 200 km
Mantle (semisolid asthenosphere and lower mantle)	Hot, dense, semisolid	More iron, magnesium, and calcium than the crust	500°C near crust to 4000°C near core	More dense than crust, less dense than solid upper mantle	2,900 km
Outer core	Thick liquid	Iron, nickel	4400°C to 6100°C	More dense than the mantle	2,200 km
Inner core	Solid	Iron, nickel	7000°C	Most dense layer	1,250 km in diameter

"Wegener" Questions

1. What evidence did Wegener use to support his idea of continental drift?

   Evidence of glacier scraping in what is now a desert near the equator; evidence of coal deposits in Arctic areas, though coal forms from tropical forests.

2. What did other scientists say about Wegener's ideas?

   They said that Wegener was a meteorologist who didn't know what he was talking about. They said that Earth is rock, and the continents can't just float around.

3. How did other scientists explain why the continents seemed to fit together?

   Many geologists thought that Earth used to be a larger globe of hot gases, and that it contracted as it cooled.

Plate Boundaries

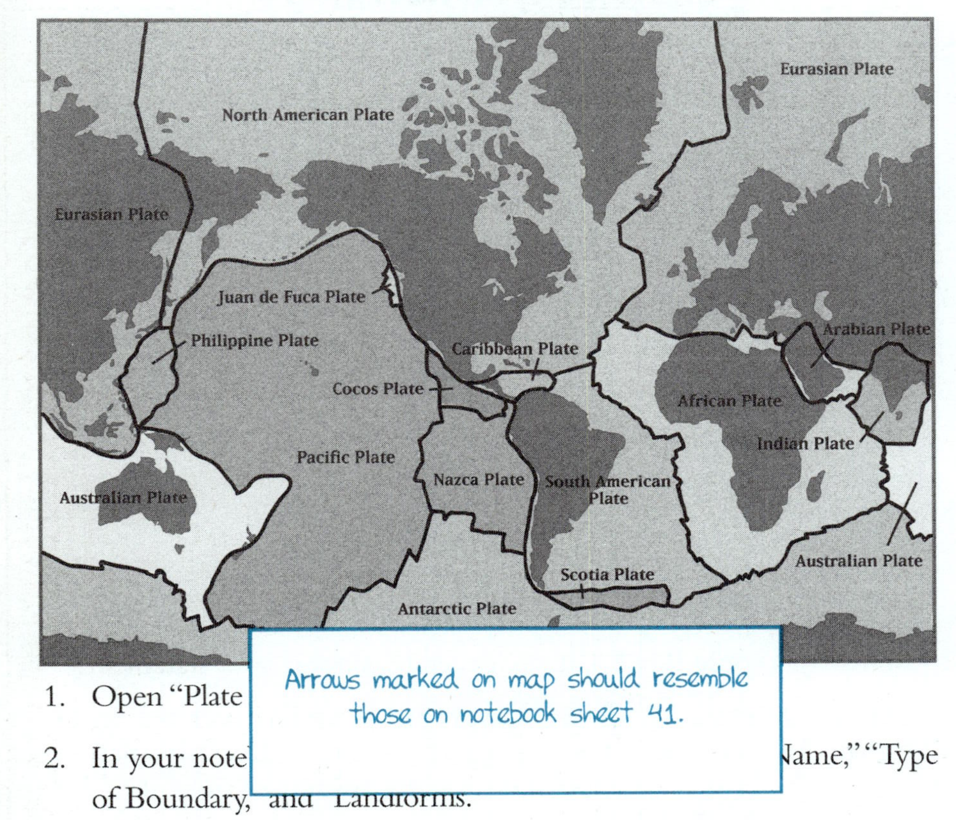

Arrows marked on map should resemble those on notebook sheet 41.

1. Open "Plate
2. In your note Name," "Type of Boundary," and "Landforms."
3. Pick five sites from the possible list (you can click on each site to visit it) and record them in the "Name" column.
4. Using the information in Google Earth™, record the type of plate boundary (divergent, transform, or convergent) in the "Type of Boundary" column.
5. For each location, mark and label the map above for that site, and draw arrows to show how the plates are moving relative to each other.
6. Record your observations about the landforms and other features you observe at that site in the "Landforms" column.

Mountain Types

Type	How it forms	Drawing	Example
Fold	Slow compression caused by converging plates.		Central Appalachian Mountains Little Stony Man, Shenandoah Medicine Bow Peak, Rocky Mountains Southern and Rocky Mountains
Fault block	Breaking of the crust either by compression or pulling apart (tension). The movement happens fast enough to break, rather than fold, the crust.		Grand Tetons, WY Sierra Nevada Basin and Range Province, CA. NV, UT Tioga Pass, Yosemite NP
Dome	Magma pushing up from below uplifts the rocks. Erosion can expose the hardened magma dome.		Half Dome, Yosemite NP Henry Mountains, UT
Volcano	Volcanic eruption.		Mt. Hood, OR Mount St. Helens Axial Seamount on the Juan de Fuca Ridge Mauna Kea
Plateau	Uplift of a relatively flat area next to a mountain that is being created by folding.		Himalaya Mountains Colorado Plateau

Field-Trip Notes

Mile 5.7 *Signal Knob Overlook*	metabasalt, sedimentary (sandstone and conglomerate) Folded during Ordovician
Mile 21 *Hogback Overlook*	gneiss
Mile 33.1 *Hazel Mountain Overlook*	Weathering broke the rocks apart. gneiss
Mile 35.1 *Pinnacles Overlook*	granite and basalt The granite came first; it had to be there for the basalt dike to cut across it.
Mile 38.6 *Stony Man Overlook*	phyllite Answers may vary for weathering and erosion.
Mile 39.1 *Little Stony Man Trail*	volcanic breccia, granite, and gneiss This looks like gneiss because it is foliated.
Mile 49 *Franklin Cliffs Overlook*	metabasaltic breccia The breccia formed when a lava flow picked up pieces of the surrounding rock. Later it was metamorphosed. The gneiss is from somewhere else, used here for construction.
Mile 51.2 *Big Meadows area, Blackrock*	Reasons for the fault may vary.

Grand Canyon Revisited A

Site 1

1. What type of rock is this? Cite evidence to support your statement.
   Limestone. It contains fossils of sea organisms, and it fizzes when acid is dropped on it.
2. What layer could it be? Cite evidence to support your statement.
   It must be the Toroweap Limestone, because you can see a sandstone layer just below it.
3. How did this layer form?
   It formed on the ocean floor, when calcium carbonate from organisms settled over time.
4. How did this layer with sea fossils get to an elevation of 2,116 meters?
   Uplift occurred after the layers formed. Movement of tectonic plates raised this rock high above where it formed.

Site 2

1. What type of rock is this? Cite evidence to support your statement.
   Sandstone. You can see the sand particles, and it does not fizz when acid is dropped on it.
2. What layer could it be? Cite evidence to support your statement.
   Coconino Sandstone. It is light-colored, and it appears near the top of the canyon. This layer forms the ring we observed in Investigation 1.
3. How did this layer form?
   Sand dunes were buried and turned into sandstone when a matrix formed between the sand grains.

Site 3

1. What type of rock is this? Cite evidence to support your statement.
   Shale. It is dark and smooth, and it does not fizz when acid is dropped on it.
2. What layer could it be? Cite evidence to support your statement.
   Hermit Shale, the only shale layer we've studied at the Grand Canyon.

Grand Canyon Revisited B

Site 3 (continued)

3. How did this layer form?

   Particles of clay sank to the bottom of a slow-moving swampy area. Over time, the clay was cemented together.

Site 4

1. What type of rock is this? Cite evidence to support your statement.

   The dark rock looks like it might be metamorphic, maybe a schist because of the foliation and large crystals. The pink rock is probably granite.

2. How does this type of rock form?

   Granite forms beneath Earth's surface, cooling slowly and forming large crystals. Schist changed from a preexisting rock buried deep in Earth and changed by heat and pressure over time.

3. How might this rock have ended up at the Grand Canyon?

   Magma in the crust cooled slowly to form the granite. It may have been forced or intruded into the schist when it was below the surface. Uplift brought it toward the surface, and weathering and erosion uncovered it. Later, the sediments that formed the sedimentary rocks covered the schist and granite.

Site 5

1. What type of rock is this? Cite evidence to support your statement.

   Basalt. It is dark gray and has tiny crystals, and it does not fizz when acid is dropped on it.

2. How does this type of rock form?

   It forms when lava cools very quickly. We know it cooled quickly because the crystals are so small.

3. How might this rock have ended up at the Grand Canyon?

   Maybe a volcano was here, or the rock could have formed somewhere else and ended up here by erosion and deposition.

This page is intentionally blank.